Biochemistry of Differentiation

Biochemistry
of
Differentiation

Charles A. Pasternak

Department of Biochemistry,
University of Oxford

WILEY-INTERSCIENCE
a division of John Wiley & Sons Ltd.
LONDON   NEW YORK   SYDNEY   TORONTO

Library of Congress Catalog card No. 74-140288

ISBN 0 471 66900 8

Printed in Great Britain
By Unwin Brothers Limited
The Gresham Press, Old Woking, Surrey, England
A member of the Staples Printing Group

Preface

It is generally accepted that biological phenomena are ultimately explicable in biochemical terms. For example, it is known that differences in species (whether animal, plant or microbe) are due to the presence of discrete enzymes catalysing specific reactions. The mechanism whereby enzymes are synthesized is by the expression of genes, made up of deoxyribonucleic acid (DNA). This concept explains how different species 'breed true', since DNA has been shown to replicate itself exactly.

Multicellular organisms of a particular species, however, produce strikingly diverse organs such as brain and liver in animals, or flower and roots in plants. Is this difference yet explicable in biochemical terms, and if so, what is the mechanism causing it? It is with these two questions that the present work is concerned.

To some, it may seem premature to attempt to present a biochemical analysis of differentiation. The subject is only just beginning to be systematically tackled at this level, and the intrinsic mechanism is not yet known. But the very fact that it is an active field of research with few established concepts makes it a somewhat confusing one to the student of biology to follow. In order to try and help him understand the current trends, a biochemical basis of differentiation is here presented. Around it are discussed some of the experimental approaches currently being used. If at the end the reader is left with little more than a greater clarity of questions, the aim may yet have been achieved. For it is often the question rather than the answer that ultimately solves the major problems of biology.

This book is based on a series of eight lectures given to third year undergraduate biochemistry and medical students at Oxford. Certain chapters are appreciably shorter than others; this is because the corresponding lecture included a film; some appropriate titles are listed in the bibliography. In order to make the subject matter readily understandable by chemist and biologist alike, a rather basic knowledge only of biochemistry has been assumed. A short bibliography, more illustrative than exhaustive, is appended to each chapter. In this, the reader will find references to some of the topics discussed in the text, as well as to matters which for reasons of space could not adequately be dealt with in the text. Review articles, sometimes several to one topic, have been cited wherever possible, in order to aid the student with limited library facilities. Where a figure is taken from an original publication, the reference is appended or given at the end of the chapter. Review articles inevitably overlap several topics, and a reference at the end of one chapter is likely to contain material pertinent to another; in order to avoid repetition, a single entry has generally been made, at the most appropriate point. For the teacher and student engaged in course work, a number of suggested questions relevant to each chapter is presented at the end of the book.

In the preparation of the course, it seemed sensible first to define a biochemical basis of differentiation (Chapter 1), next to test its applicability in those systems of microbes, plants and animals that have proved amenable to experimental examination (Chapters 2–6) and finally to consider possible mechanisms underlying differentiation (Chapters 7 and 8). However there may be some who prefer to put hypothesis before experiment, and in that sense the last two chapters may be read before the others without much loss of continuity.

Acknowledgements

I am indebted to many colleagues for helpful discussions, and especially to J. J. M. Bergeron, R. J. Ellis, J. B. Gurdon, R. E. Handschumacher, Henry Harris, Sir H. A. Krebs, E. S. Lennox, B. C. Loughman, J. Mandelstam, P. C. Newell, D. S. Parsons, R. R. Porter and J. R. Tata for reading all or parts of the manuscript.

To J. R. Baker, D. Kay, H. O. Halvorson, V. Vinter, S. C. Warren, M. Sussman, J. T. Bonner, E. C. Cantino, K. Esau, J. van Overbeek, D. H. Northcote, C. J. Avers, C. H. Waddington, E. N. Harvey, A. Monroy, J. B. Gurdon, G. W. Brown, D. G. Walker, W. J. Rutter, J. Gross, H. E. Huxley, L. F. Leloir, C. L. Markert, P. G. W. Plagemann, A. S. Parkes, H. H. Ross and the Illinois Natural History Survey, W. Bartley, W. Beermann and C. Pelling, J. R. Tata, W. E. Knox, G. Weber, T. W. Sneider, H. C. Pitot, H. Harris, H. Green, F. C. Steward, B. S. Tyler, D. D. Brown, J. Bonner and B. J. McCarthy I am grateful for permission to reproduce or adapt original photographs, figures or other material.

C.A.P.

Contents

CHAPTER 1

Introduction

1-1 Definitions

(a) *Nature of Differentiation*

During the development of an animal from the fertilized egg, all *differentiated* structures such as liver, heart, brain, eye appear. They are easily recognized by their *size*, their *shape*, their *structure* and their *function*. The same is true of the emergence of roots, stalk, leaves and flowers from the fertilized ovum in plants. Differentiation, then, might be defined as the development of unique clusters of cells called organs. But single cells themselves develop from one type to another during the life of an animal, as in the formation of enucleated red cells from nucleated precursor cells. The development of a sporulating cell from a normal vegetative cell is an example of a similar process in microbes. In each case the new cell is recognizably different from the old in terms of structure and function. Development of altered structure or function, then, may be a sufficient criterion of differentiation.

However, cells alter their structure and function in a manner which falls outside this definition. For example, some photosynthetic bacteria such as *Rhodopseudomonas spheroides* can live aerobically in the dark or anaerobically in the light. The change from dark to light environment is accompanied by an alteration in metabolism and by the appearance of the structures (chromatophores) necessary for photosynthesis. Less dramatic alterations occur when bacterial cells, such as *Escherichia coli*, growing on glucose adapt to growing on lactose instead. Animal cells also adapt to altered environmental conditions. Muscle cells, for example, normally metabolize glucose aerobically to carbon dioxide and water; during violent exercise, however, the oxygen supply becomes limited and anaerobic metabolism to lactic acid takes place instead. In fact the ability to adapt to the environment is a characteristic feature of most living cells.

(b) Differentiation versus Adaptation

One must therefore distinguish between *adaptation* and *differentiation*. The basic difference lies in the reversibility of adaptive phenomena. Thus *R. spheroides* reverts to aerobic metabolism when the light is switched off, *E. coli* loses the ability to degrade lactose when returned to glucose medium and muscle cells revert to aerobic degradation of glucose when the oxygen supply is restored. The reversions are not instantaneous but are delayed by as long as it takes one set of enzymes to be replaced by another, or to be modified in other ways. Unlike adaptive phenomena, differentiation is generally irreversible and stable. Liver cells cannot change back to an ovum, or red blood cells to nucleated precursor cells, any more than leaf or root cells can revert to seeds (some exceptions, albeit rather unphysiological, to this general rule are discussed in Section 1 (*c*) below and in Section 7-1). In other words, *differentiation is the stable development of altered structure and function.*

(c) Commitment

But spores do change back into normal dividing cells by the process of germination (Figure 1-1). Should spore formation, then,

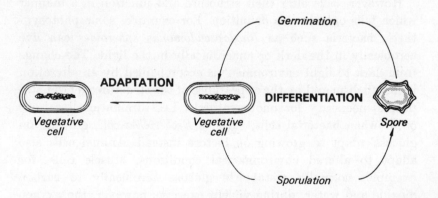

Figure 1-1 Adaptation *versus* differentiation in microbes

be considered a process of adaptation rather than of differentiation? Either case could be argued but for this treatment it will be included under differentiation. The point is this. One of the stimuli for the formation of bacterial spores (sporulation) is the exhaustion of nutrients. If nutrients are restored to the culture immediately after depletion, spore formation does not take place; the culture is in the reversible, adaptive state. But if nutrients are added some time after spore formation has already begun, sporulation continues despite the presence of nutrients. In other words, cells are now *committed* to sporulation. Likewise spores become committed to germination on the return to more favourable conditions. It is the presence of a commitment stage, also referred to as *determination* during embryological development, which allows one to distinguish between differentiation and adaptation. In certain cases this is not quite true. Slime mould cells (Section 2-2), for example, become committed to fruiting body formation *after* the plasmodium has already differentiated into presumptive spores and stalk cells. One could, of course, describe the pre-commitment phase of slime mould development as adaptation rather than as differentiation, but such restrictions on the use of generally accepted terms seem to be pedantic rather than useful.

It might be argued that the differentiated state is stabilized only by the continued presence of substances, as yet unidentified, within or around cells in a way that adaptive phenomena are maintained. Moreover, there may be temporary commitment to a particular function during the reversible, adaptive process. Insofar as the processes of differentiation and adaptation are each manifestations of an altered synthesis of specific proteins (see below), the mechanism is clearly similar. The difference lies in the fact that the signals for differentiation are somehow maintained and inherited in dividing cells in a manner distinct from that occurring during adaptation. The *stable* development of altered structure and function is a characteristic feature of differentiation.

(d) *Initiation and Induction*

The agents that trigger off a chain of developmental events have been referred to as *initiators* and their action as *initiation*. An example of initiation in the case of slime mould development is presented in Chapter 2. In other situations, such as in the development of morphologically distinct structures during embryonic growth, the terms *inducer* and *induction* are used. Some systems in which induction has been studied are described in Section 4-1(*c*).

1-2 Biochemical Basis of Differentiation

We come now to the central theme of this treatise. What are the essential biochemical features that account for the variations in size, shape, structure and function of differentiated cells and organs? It must at once be admitted that in many cases the answer is just not known. For example, there is as yet no idea about what determines the size and shape of organs. Ignorance concerning the limitation of cellular growth is reflected in our inability to explain the phenomenon of cancer (see Chapter 6) in biochemical terms. As regards the determination of structure and function, on the other hand, there is rather a surfeit of possible factors to consider and the difficulty is to decide which is the primary one.

(a) Comparison between Muscle and Liver

Compare a muscle cell with a liver cell. In structure a muscle cell is recognized by its association with strands of contractile protein, the actomyosin fibrils (Plate 1, facing p. 4). The liver cell does not possess these, being largely concerned with metabolic activity such as the degradation and resynthesis of proteins, fats and carbohydrates. It is, in contrast, rich in endoplasmic reticulum and ribosomes (Plate II).

Muscle and liver cells contain mitochondria for converting the energy of foodstuffs into, in the one case, mainly mechanical energy and in the other, chemical energy. Other differences exist. Liver mitochondria synthesize urea whereas muscle ones do not. Cells of each organ contain glycogen but the amounts are different. So are the concentrations of glucose, which are determined mainly by the type of membrane which surrounds the cell; muscle membranes are sensitive to insulin which causes increased absorption of glucose; liver cells are insensitive to the hormone. Liver cells store compounds such as vitamin A, muscle cells contain creatine,

PLATE I

Electron micrograph of leg muscle (mouse), magnified ×30,000. Note the mitochondria (M) lying close to the actomyosin fibrils (F). Contraction of the fibrils is shown in Figure 4-5 (Taken by Dr. J. R. Baker)

PLATE II

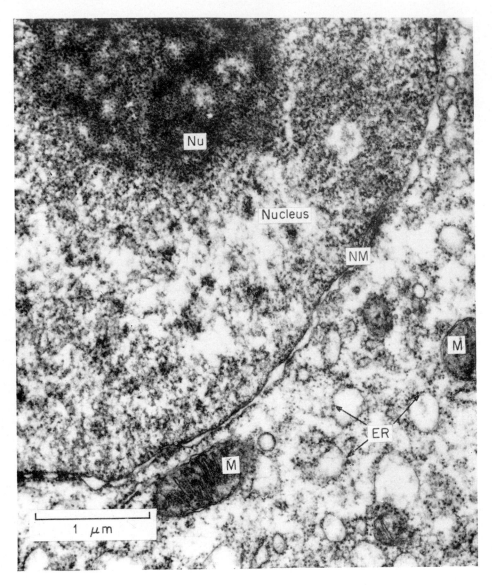

Electron micrograph showing part of a liver cell (mouse), magnified ×30,000. Note the nucleus, containing nucleolus (Nu) and bounded by the nuclear membrane (NM), mitochondria (M) and endoplasmic reticulum (ER) studded with ribosomes (Taken by Dr. J. R. Baker)

and so on and so forth. The constituents vary not only in kind and in amount but also in their properties. Liver phosphorylase is not the same enzyme as muscle phosphorylase and it responds to hormones to which muscle is insensitive. What underlies all these variations? The thesis of this book is that it is proteins, specific with regard to enzymic or structural function, which are immediately responsible for the differences just enumerated.

(b) Proteins as Basic Controlling Units

That specific enzymes are reponsible for synthesizing distinct compounds such as glycogen or phospholipids or urea or creatine will not be questioned. Nor can it be disputed that the presence of a protein such as actomyosin or haemoglobin lends to the cells in which it occurs the specific property of muscle contraction or oxygen transport. But that the rate of glycolysis of a liver cell, as opposed to that of a muscle cell, is controlled by the type or amount of glycolytic enzymes it contains is not apparent. Indeed, it may be argued that it is the amount of glucose, or ATP, or citrate, or other inhibitor or activator that controls the rate. Where enzyme is not saturated by substrate, activator or inhibitor, this is certainly true. But the concentration of ATP or citrate is itself dependent on the enzymes that synthesize and degrade it, just as the concentration of glucose is dependent on the nature of the membrane—containing structural and enzymic proteins as well as carbohydrate and phospholipid—which surrounds the cell. So that one may, for simplicity, accept the situation outlined in Figure 1-2. By cellular organization is meant

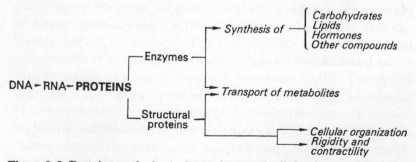

Figure 1-2 Proteins as the basic determinants of cellular form and function

the membranous system (the 'cytoskeleton' of Chapter 5-1(e)), predominantly protein and phospholipid, that delineates the surface of the cell and the organelles such as nucleus and mitochondria within it.

It should be noted that the distinction between structural and enzymic proteins is somewhat arbitrary. Haemoglobin is not an enzyme; yet it combines with oxygen in a manner very characteristic of enzymes. Actomyosin is the structural unit of muscle fibres but it also possesses the ability to degrade ATP. And the structural proteins of mitochondria and chloroplasts catalyse some of the energy transformations characteristic of these organelles.

(c) *DNA and RNA*

The reader may criticize Figure 1-2 and say that, since proteins are synthesized from DNA through the agency of RNA (Figure 8-1), it is the content of DNA which really characterizes differentiated cells. This is certainly a valid suggestion and the view that differentiation is a mutational event causing the acquisition, loss or alteration of existing genes is an attractive one. It has, however, become untenable as a result of some important experiments to be described in Section 7-1.

The role of RNA in differentiation cannot be so easily settled. It depends on whether one believes that protein synthesis is controlled at transcription (in which case the synthesis of specific RNA molecules coding for particular proteins will accompany differentiation) or at translation (in which case the RNA content of differentiating cells may, like DNA, remain unaltered). Since it is as yet not possible to recognize an RNA molecule as specifying one sort of protein rather than another, this question cannot be answered. Current speculation on the problem of transcriptional *versus* translational control of protein synthesis is summarized in Section 8-1. In the meantime it will prove more profitable to take proteins as the basic units and examine to what extent they can account for the variations of structure and function apparent in differentiated cells. The examples, which must be selective since differentiated structures comprise practically all living matter, will be taken from microbes, plants and animals wherever an appropriate situation has been critically studied. In one case, an analysis of nucleic acid synthesis will be presented. This is in the development of fertilized animal eggs (Chapter 3) in which the synthesis of DNA and RNA is not closely linked to that of proteins.

1-3 Concluding Remarks: Differentiation in Terms of Specific Proteins

The pattern of proteins within a cell is a reflexion of the relative concentration of individual molecules; alteration in the *rate of protein synthesis* (from zero upwards) can account for variation both in *kind* and in *amount* of protein present. In the case of enzymes (and certain proteins such as haemoglobin), overall function depends not only on the kind and amount of enzyme present but also on its *activity*. Activity is determined by many factors, including the presence of specific effectors such as a substrate, product, hormone or some other molecule. The activity of enzymes may vary from one cell type to another as discussed in Section 4-2(*b*). It is not always easy, when considering the proteins of differentiated cells, to distinguish between altered *synthesis* and altered *activity* of enzymes. An attempt to do so wherever possible will be made in the chapters that follow. It will transpire in a discussion of isoenzymes (Section 4-2(*b*)) that altered activity is itself generally a consequence of altered synthesis of particular subunits.

1-4 Summary

Differentiation may be defined as the *stable development* of *altered structure* and *function*. Development may involve the formation of whole organs of specific *size* and shape. The *stimulus* for differentiation is sometimes environmental, as in the formation of bacterial spores. In that case, it can be distinguished from *adaptation* to the environment by the fact that, once the cells become *committed* to differentiation, the process is generally irreversible and stable. Agents that cause differentiation are known as *initiators* or *inducers* and their action as *initiation* or *induction*.

The biochemical basis of differentiation lies in the production of specific proteins. These function either as *structural* units, such as actomyosin of muscle, or as *enzymes* catalysing the synthesis and transport of a variety of molecules.

The pattern of proteins in differentiated cells may be varied by an alteration in their rate of *synthesis*. In the case of enzymes, altered *activity* likewise leads to variation in cellular function.

Selected Bibliography

Barth, L. J. (1964). *Development and Selected Topics*. Addison-Wesley, Reading, Mass.

Beermann, W. and coworkers (1966). *Cell Differentiation and Morphogenesis*. North-Holland, Amsterdam

Brachet, J. (1960). *The Biochemistry of Development*. Pergamon Press, Oxford

Ebert, J. D. (1965). *Interacting Systems in Development*. Holt, Rinehart and Winston, New York

Fogg, G. E. (Ed.) (1963). Cell Differentiation. *Symp. Soc. Exp. Biol.*, **17**

Grobstein, C. and coworkers (1961). Differentiation of Vertebrate Cells. In *The Cell* (Ed. J. Brachet and A. E. Mirsky), Academic Press, London–New York, Vol. 1, p. 437

Mazia, D. and A. Tyler (Ed.) (1963). *General Physiology of Cell Specialization*. McGraw-Hill, New York

Sussman, M. (1964). *Growth & Development*. Prentice-Hall, Englewood Cliffs, New Jersey

Telfer, W. H. and D. Kennedy (1965). *The Biology of Organisms*. John Wiley & Sons, New York

Trinkhaus, J. P. (1970). *Cells into Organs*. Prentice-Hall International, Hemel Hempstead

Waddington, C. H. (1966). *Principles of Development and Differentiation*. Macmillan, London

Whittaker, J. R. (1968). *Cellular Differentiation*. Prentice-Hall International, Hemel Hempstead

Wright, B. E. (1964). The Biochemistry of Morphogenesis. In *Comparative Biochemistry* (Ed. M. Florkin and H. S. Mason), Academic Press, London–New York, Vol. 6, p. 1

CHAPTER 2

Differentiation in microbes and plants

In this chapter the development of bacterial spores, slime moulds and some differentiated cells of plants will be examined. The relevance of the first two is emphasized by the fact that they are currently under vigorous analysis by several groups interested in differentiation.

2-1 Bacterial Spores

(a) Life Cycle: Sporulation and Germination

Several types of bacteria, mostly bacilli but including other microbes as well, possess the ability to remain dormant as *spores* for very long periods of time.   The seeds of plants are an example of this phenomenon in multicellular organisms. The stimulus for spore formation, or *sporulation* (also known as sporogenesis) is generally the deterioration of environmental conditions, such as lack of sufficient nutrients, to the point where normal growth and cell division stop. In bacilli, nuclear bodies of the surviving cells begin to reorganize themselves so that one complete genome becomes encased at one end of the cell, which begins to synthesize a thicker outer wall and loses water. Eventually a mature spore is liberated and the rest of the cell disintegrates. Sporulation may be complete within eight hours, depending on the temperature and other factors (Figure 2-1). Spores are characterized by their resistance to external conditions: extremes of temperature, ionic environment, humidity, ultraviolet light, x-rays and other radiations. The stimulus for the return to vegetative life, or *germination*, is the restoration of favourable environmental conditions. Then spores are rehydrated, the outer wall is shed and a vegetative cell,

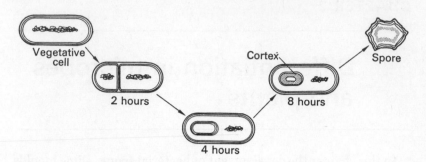

Figure 2-1 Stages in the sporulation of a typical bacillus; in this instance, maturation took some ten hours to complete. Redrawn from an original plate of Dr. D. Kay

capable of normal metabolism and division, is released. Some biochemical changes that may account for the behaviour of spores will now be considered.

(b) Heat and Radiation Resistance

One of the most puzzling aspects of spores is how their enzymes are able to avoid heat denaturation and resultant loss of activity. The situation is somewhat similar to that of thermophilic bacteria which grow in sulphur springs at 70–80°C. The dehydrated nature of spores is undoubtedly a major factor, though how dehydration is brought about is not known. The thick outer wall may play a part in maintaining dehydration; but the wall is unlikely to act directly as an insulator, since spores can withstand heating to 100°C for several hours.

A characteristic component of bacterial spores, absent in vegetative cells, is dipicolinic acid (Figure 2-2). Although its functional

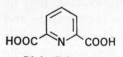

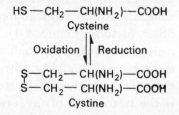

Dipicolinic acid

Figure 2-2 Some typical constituents of bacterial spores

relation to heat resistance is not known, the amount of dipicolinic acid which appears in sporulating cells seems to be temporally connected with the resistance to heat inactivation (Figure 2-3). It

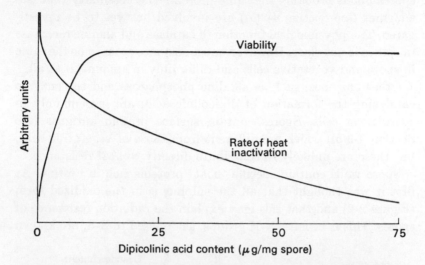

Figure 2-3 Development of heat resistance and viability of *B. cereus* spores with increasing dipicolinic acid content. From Church and Halvorson (1959). *Nature,* **183,** 124

is not yet possible to account for the appearance of dipicolinic acid in terms of the synthesis or activation of specific enzymes because the biosynthetic pathway has only recently been elucidated.

Another component of bacilli that increases during sporulation is the content of calcium and this, too, has been implicated in heat resistance. Thus the content of calcium increases simultaneously with that of dipicolinic acid (with which it probably forms a complex) and sporulation in the absence of adequate amounts of calcium in the medium leads to heat-sensitive spores. Copper ions are also accumulated during sporulation and have been implicated in heat resistance.

Several investigators have studied the actual enzymes found in spores. Catalase, for example, exists in two distinct forms, a heat labile and a heat stable one. The heat stable variety which occurs in spores appears to have considerably fewer hydrogen bonds than the heat labile one of vegetative cells, in accord with the fact that rupture of hydrogen bonds is one of the main consequences of heat denaturation. Alanine racemase also has altered properties in spores but is probably the same protein. The possibility that isoenzymes (see Section 4-2(b)) are involved has yet to be investigated. The physiological function of catalase and alanine racemase in spores is not clear. However, most enzymes seem to be the same in spore and vegetative cells and differ only in amount, if at all.

Other enzymes, such as alkaline phosphatase and the enzyme catalysing the formation of dipicolinic acid, are present only in sporulating cells. Spores contain surface protein antigens (see Section 4-3(b)) which are different from those of vegetative cells, but these are unlikely to contribute directly to heat resistance.

Spore walls contain specific 'coat' proteins rich in cystine. At first it was thought that all the sulphur is in the oxidized form (Figure 2-2) and that this may explain the radiation resistance of spores. This is because S–H groups, as opposed to S–S, are known

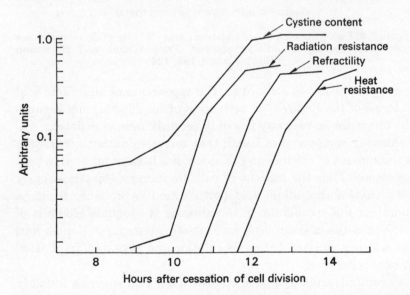

Figure 2-4 Development of radiation resistance in *B. cereus* spores with increasing cystine content. The increase in refactility corresponds to the formation of the cortex (see Figure 2-1). From Vinter (1961). *Nature,* **189**, 589

to be rather sensitive to the presence of free radicals which are formed when cells absorb radiations. However it now appears that S–H groups *are* present in spores to the same extent as in vegetative cells but that they are 'buried' within the proteins as a result of an altered secondary structure. What is clear is that the increase in cystine content during sporulation coincides with the emergence of radiation resistance, whereas heat resistance, for example, becomes manifest only some two hours later (Figure 2-4).

One must conclude that although several proteins have been found to alter during sporulation, a completely biochemical explanation of the behaviour of spores is not yet possible.

(c) Commitment during Sporulation

The sequential appearance of specific enzymes and properties such as heat and radiation resistance during the formation of bacterial spores (Figure 2-5) has enabled one to study some aspects

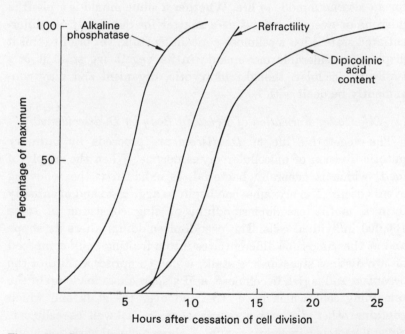
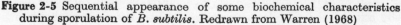

Figure 2-5 Sequential appearance of some biochemical characteristics during sporulation of *B. subtilis*. Redrawn from Warren (1968)

of commitment under defined conditions. Thus, if the antibiotic
actinomycin D (see Section 8-1(b)) is added at the time that
sporulation is induced by removal of nutrients, none of the events
referred to in Figure 2-5 occurs. If added two hours later, alkaline
phosphatase appears but no subsequent event occurs. If added
three hours later, some spores appear as well as alkaline phosphatase (Figure 8-6), but they contain no dipicolinic acid. In other
words, bacteria seem to become *committed* to synthesizing alkaline
phosphatase before they are committed to forming spores. Strictly
speaking all one cay say is that the expression of alkaline phos-
phosphatase is already determined. 'Potentiality' perhaps de-
scribes the situation better than commitment.

2-2 Slime Moulds

The production of slime moulds bears a resemblance to that of
bacterial spores in so far as lack of nutrient is again the stimulus
for an alternate mode of life. Whether a slime mould is a plant, a
microbe or even an animal, is a matter for debate. In its differentiated state it has a cellulose-containing stalk, yet prior to this it
displays considerable movement; in its vegetative state it is a
typical unicellular, though eukaryotic, organism and may con-
veniently be dealt with here.

(a) Life Cycle: Formation of Fruiting Body by D. discoideum

The vegetative life of *D. discoideum* proceeds by ordinary
mitotic division of unicellular myxamoebae. When the supply of
food, which is generally bacterial, is exhausted, the following
events occur. The myxamoebae begin to aggregate and eventually
form a motile pseudoplasmodium or slug consisting of some
150,000 individual cells. The pseudoplasmodium alters its shape
and at the same time differentiates into a fruiting body composed
of two distinct structures: a stalk, which comprises a third of the
organism and is rich in cellulose, and a spore mass made up of the
remaining cells which have streamed over the stalk and which
contains other distinctive polysaccharides as well as cellulose.
Several proteins, including enzymes, show regional variation along
the axis of the pseudoplasmodium. When the fruiting body has

matured, the spore mass disintegrates and releases the constituent spores; these eventually germinate into myxamoebae which undergo normal division once more. The whole sequence of events up to fruiting body takes some 20–50 hours (depending on the conditions) to complete (Figure 2-6). At no stage after aggregation do any cells divide.

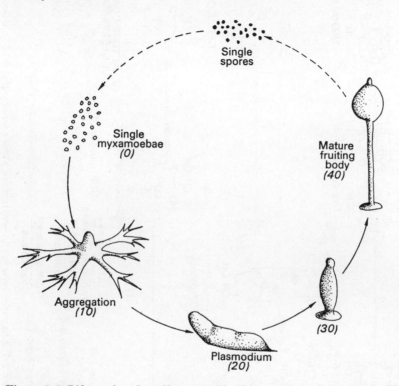

Figure 2-6 Life cycle of a slime mould (*Dictyostelium discoideum*). The numbers in the figure are hours after cessation of cell division. From Sussmann (1967)

(b) *Polysaccharide Synthesis*

One of the interesting aspects of slime mould differentiation is to determine where and when polysaccharide synthesis takes place, since specific polysaccharides occur in the differentiated stalk and spore cells. Apart from cellulose, the fruiting body of *D. discoideum* appears to contain a glycogen-like polymer of glucose,

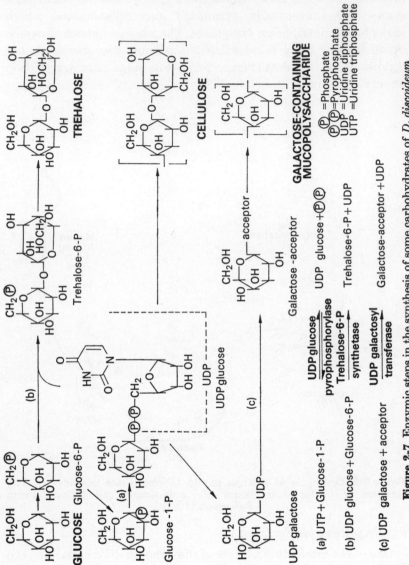

Figure 2-7 Enzymic steps in the synthesis of some carbohydrates of *D. discoideum*

trehalose (a disaccharide of glucose) and a mucopolysaccharide made up of galactose, galactosamine and galacturonic acid. The last two are localized in the spore cells. A key enzyme in the synthesis of the mucopolysaccharide is uridine diphosphate (UDP) galactosyl transferase (Figure 2-7). The stage in development at which it, and the polysaccharide which it synthesizes, appear are shown in Figure 2-8. Addition of inhibitors of protein synthesis

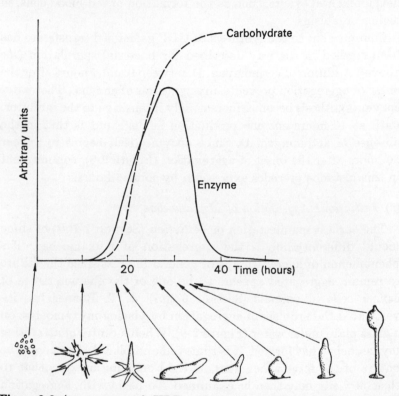

Figure 2-8 Appearance of UDP galactosyl transferase (enzyme) and galactose-containing mucopolysaccharide (carbohydrate) during development of *D. discoideum*. From Sussman (1967)

has shown that formation of new enzyme, rather than activation of inactive precursor, is involved. Note that enzymic activity decreases after approximately 30 hours whereas the amount of polysaccharide remains constant. The synthesis of two other enzymes (Figure 2-7) has been followed. Trehalose-6-phosphate synthetase appears at 7 hours and UDP glucose pyrophosphorylase (which catalyses the initial step in the synthesis of several polysaccharides) at 12 hours.

Since no cells divide during slime mould differentiation, the process exemplifies the emergence of new structure and function within *existing* cells, just as in bacterial sporulation and in many developmental events, such as the formation of red blood cells, in higher organisms.

Commitment to the synthesis of UDP galactosyl transferase has been studied in the way described for bacterial sporulation (see above). Addition of actinomycin D between 0 and 7 hours after the onset of aggregation prevents any synthesis of enzyme. Thereafter enzyme synthesis becomes increasingly insensitive to the inhibitor, until at 15 hours enzyme production is the same as that in the absence of actinomycin D. Since enzyme itself begins to appear 20 hours after the onset of aggregation (Figure 2-8), commitment in this instance precedes expression by some 5 hours.

(c) Initiation: Aggregation of Myxamoebae

The earliest manifestation or *initiation* (Section 1-1(*d*)) of slime mould differentiation is the aggregation of myxamoebae. The phenomenon of aggregation is of general interest since the failure to remain aggregated appears to be one of the characteristics of cancer cells in animals (Section 6-2(*b*)). J. T. Bonner has investigated the problem of aggregation by placing myxamoebae on a glass plate under water (Figure 2-9). When a centre of attracting myxamoebae has formed, it is removed and placed elsewhere. The nature of the force which attracts the remaining myxamoebae to their new site may then be examined. To begin with, aggregation cannot be due to direct contact between cells, since the attractive force is operative up to a distance of 200 μm, whereas the diameter of a single myxamoebae is only 15 μm. Electric currents (which cause other cells, such as *Amoeba proteus* to stream to the cathode) or magnetic fields are without effect. Thin glass (120 μm), mica (100–150 μm), quartz (50–100 μm) or tantalum (12 μm) placed between the centre and the cells prevents aggregation but cellophane

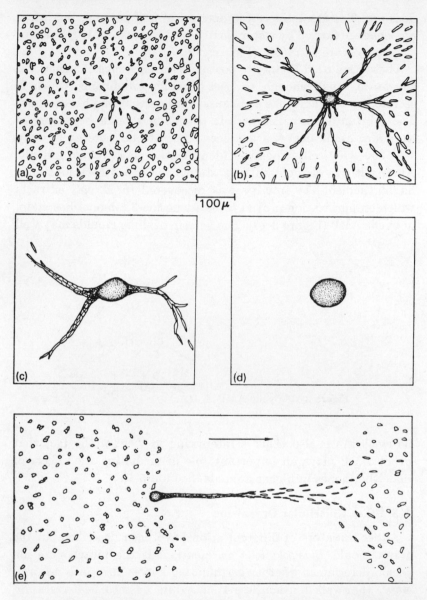

Figure 2-9 Aggregation of *D. discoideum* myxamoebae on a glass dish under water. a–d: formation of an aggregation centre in still water; e: formation of an aggregation centre under water moving in the direction of the arrow. From J. T. Bonner (1965)

2

does not. The attractive force operates round corners, but when amoebae are carried by water current past a centre, only the downstream cells are attracted (Figure 2-9).

These results led Bonner to postulate the existence of a chemical substance, small enough to pass through cellophane, which is secreted by myxamoebae and causes the aggregation phenomenon. The substance was called acrasin (the slime moulds belonging to the order Acrasiales) and has been shown to be secreted by only 1 in 300–2000 cells. Chemical identification has proved difficult. At first it was thought to be a steroid but recently a substance having acrasin-like activity was discovered in *E. coli* extracts which on purification proved to be adenosine-3,5-monophosphate, or cyclic AMP (Figure 2-10). The acrasin of slime moulds may well

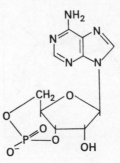

Figure 2-10 Cyclic AMP

be cyclic AMP also. This is interesting in view of the fact that cyclic AMP plays an important role in mediating the effects of many hormones in higher animals (Section 5-1(c)).

2-3 Other Unicellular Organisms

Other initiators of differentiation have been described. In the water mould *Blastocladiella emersonii*, it is the bicarbonate ion which determines whether germinating spores turn into an ordinary, thin-walled, colourless sporangium or into heat-resistant, thick-walled, brown sporangium (Figure 2-11). In developing embryonic chick rudiments (Section 4-1(c)), another nucleotide has been implicated as initiator. It is as well to bear in mind, how-

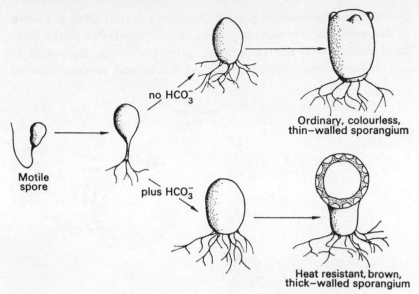

Figure 2-11 Effect of bicarbonate in determining development of the water mould *Blastocladella emersonii*. From Cantino and Lovett (1964)

ever, that the compound which is tentatively identified as an initiator may itself be formed by the production or release of other potentially active compounds, and *vice versa*.

A unicellular organism which has been much studied in relation to differentiation is the alga *Acetabularia*. Its properties will be described in Section 8-1(*b*).

2-4 Plants

(*a*) Introduction

The germination of seeds to form specialized organs such as roots, stem, leaves and fruit appears at first sight to be the most clear cut example of differentiation in plants. However, the seed itself is already made up of rather specialized cells in the way that a mature animal embryo is, and it is therefore seed formation which is the analogous developmental event. Unlike the development of an animal embryo, which is beginning to be quite extensively described in biochemical terms (Chapters 3 and 4), little is known of the maturation process in plants. Mainly this is due to the technical difficulty of analysing seed, or rather ovule, formation in developing flowers under controlled conditions. Germination of

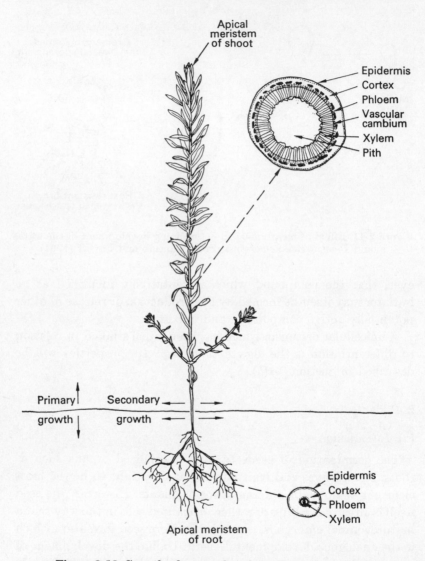

Figure 2-12 Growth of a vascular plant. From Esau (1965)

seeds, on the other hand, has been rather more widely studied, particularly in the test tube. The later stages of plant growth differ from those in animals in that undifferentiated meristematic cells, situated at the growing points of root and shoot (Figure 2-12) continue to divide and to give rise to differentiated adult cells throughout the life of the plant. In animals, adult morphogenesis is confined to certain tissues such as those involved in erythro-poiesis (Section 4-3(a)). The elongation of root, shoot and other tissues is known as primary growth and precedes the thickening, or secondary growth of these tissues. Two structures which are characteristic of adult plant cells, the chloroplast and the cell wall, will briefly be considered. Finally, some biochemical aspects of the development of certain specialized plant cells will be described.

(b) Germination

One of the main biochemical events during seed germination is the breakdown of the stores of carbohydrate, fat and protein

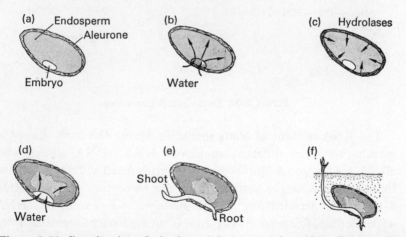

Figure 2-13 Germination of a barley seed. As water from the soil enters the seed (b), the plant hormone gibberellin is released and diffuses into the aleurone layer where it stimulates production of hydrolases such as α-amylase (c). This leads to the breakdown of the food stores in the endo-sperm. As more water enters, the formation of auxin and other hormones (d) leads to the appearance of root and shoot ((e) and (f)). From Van Overbeek
(1968)

which are laid down during seed formation (Figure 2-13). Both activation and synthesis of enzymes are involved. The degradation of starch in germinating barley seeds, for example, is catalysed by two enzymes: β-amylase, which is present in the seed, and is activated in some way on germination; and α-amylase, which appears only during germination, and which has been elegantly shown by Varner to involve the synthesis of new enzyme molecules. The formation of α-amylase and other enzymes is stimulated by the release of a plant hormone known as a gibberellin (Figure 2-14). The effect of hormones in increasing protein synthesis is a typical one and is seen in animals (Section 5-1(*b*)) as well as plants.

AUXINS

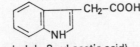

(e.g. indol–3–yl acetic acid)

KININS

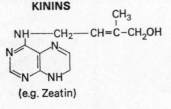

(e.g. Zeatin)

GIBBERELLINS

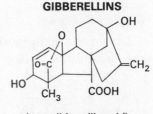

(e.g. gibberellic acid)

Figure 2-14 Some plant hormones

The RNA content of seeds increases during the early stages of germination. The different species of RNA (tRNA, dRNA and rRNA, see Section 3-3(*a*)) begin to be synthesized at different times, indicating a selective activation of the respective genes (Section 8-1(*b*)). In germinating wheat embryo, for example, tRNA appears first, followed closely by rRNA and only sometime later by dRNA. RNA synthesis is stimulated by gibberellin; thus the induced synthesis of α-amylase and other enzymes is probably due to increased availability of mRNA templates.

Some of the proteins that are stored in seeds are rather specific. Pea seeds, for example, contain two globulins not found elsewhere in the plant. An attempt to study their synthesis in the test tube has been made by J. Bonner. He prepared extracts of various pea plant organs, such as flower, cotyledon, roots, apical bud and leaves and incubated them with a radioactive amino acid. All extracts incorporated radioactivity into protein. However, when the proteins were fractionated by an immunological technique (Section 4-3(b)) designed to separate the specific pea seed globulins, only flower and cotyledon were found to have synthesized any; this is the result expected from the known distribution of the proteins. In principle, this type of experiment is an excellent example of the synthesis of specific proteins by differentiated tissues. So far, however, it has not been repeated by Bonner or by other workers.

Several other enzymic activities vary during germination. An interesting example is that of glutamate dehydrogenase. When the enzyme is isolated from pea embryos, it can be dissociated into some seven sub-units. The same sub-units are found in the enzyme from shoots or radicals, but the proportions in which they are present are characteristically different. The resulting enzymes are termed 'isoenzymes' (Section 4-2(b)). Peroxidase of peas also exists as isoenzymes. Again the pattern of sub-units alters as different plant organs develop. Many other examples of isoenzymes are known. Since the catalytic properties of an enzyme depend on the nature of its sub-units, variation in the proportion of sub-units can bring about quantitative alterations in metabolic pathways. The relevance of such changes to differentiation is discussed in Section 4-2(b).

(c) Development of Specialized Structures

(i) Chloroplasts. Photosynthesis—the generation of chemical energy at the expense of light—might appear to be a good example of differentiation in plants. Certainly chloroplasts, which are structures somewhat larger in size than mitochondria and which catalyse the photosynthetic process by specific enzymes, are present predominantly in leaves. But chloroplasts are sometimes also found in stems, which also contain the characteristic pigment chlorophyll, though not in roots. It may therefore be argued that the presence of chloroplasts is typical of plant cells generally (as opposed to animal or microbial cells) and that it is the *absence* of chloroplasts from particular cells that distinguishes these in the way that the absence of mitochondria is a characteristic feature of the highly differentiated animal red cell (Section 4-3(*a*)). Nevertheless, the formation of chloroplasts and their constituent enzymes and other proteins during the growth of a plant is clearly an important developmental event. Little is yet known of the way in which the characteristic proteins and other molecules are synthesized and assembled, but it is a field in which current research is active.

(ii) *Cell Walls.* Another structure unique to the plant (and microbial) kingdom as a whole is the cell wall. It is more widely distributed in adult cells even than the chloroplast, but again its formation in growing cells is a good illustration of the development of a differentiated structure. Higher plant cell walls, which surround the plasma membrane, consist of a continuous matrix in which strands of cellulose (a polymer of 1–4-β-linked glucose units) are embedded. The matrix of the wall is made up of a mixture of polysaccharides called pectin and hemicellulose. Pectin contains polymers of galacturonic acid, arabinose and galactose; hemicellulose contains polymers of xylose, glucuronic acid, glucose and mannose. The constituents of the cell wall accumulate and alter in a defined sequence. In cambial stem cells, for example, the acidity of the matrix decreases during development. This is due to two causes: first, the proportions of the constituents of pectin alter, galacturonic acid being partly replaced by galactose and arabinose; secondly, pectin formation stops whereas hemicellulose production continues. Since the components of pectin are metabolically related to those of hemicellulose by epimerization (Figure 2-15), it is possible that this stage of development results from a loss of the epimerase enzymes.

The constituent polymers of pectin and hemicellulose are synthesized within the cytoplasm and appear to be transported to the site of wall formation by small structures known as microtubules. Cellulose, on the other hand, is synthesized at the surface of the

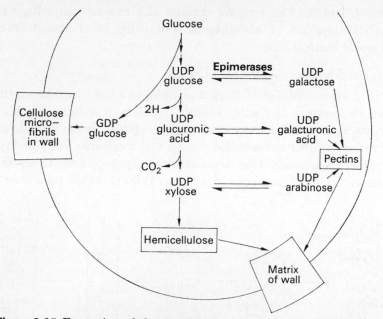

Figure 2-15 Formation of the constituents of plant cell walls (see Figure 2-7 for the synthesis of UDP glucose from glucose). From Northcote (1969a)

membrane and laid into fibrillar strands at the same site. The final stage of wall formation may involve the deposition of lignin (a polymer of phenolic derivatives). This gives the wall rigidity and eventually renders it impermeable.

In certain root cells, quinones are synthesized and are deposited in the cellulose strands; combination with protein ('tanning') leads to a hardened surface. Quinones are rather toxic and possibly their function is to prevent microbial invasion. Quinones are synthesized from aromatic amino acids by way of various phenolic intermediates (Figure 2-16), the final stage being catalysed by poly-

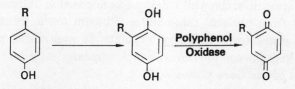

Figure 2-16 Stages in quinone formation

phenol oxidase. This enzyme appears at a characteristic stage in the development of endodermal root cells. It is absent from adjacent cortical cells.

(d) Development of Specialized Cells

(i) *Phloem and Xylem.* Together these two tissues make up the vascular system of plants. Phloem consists of a series of tubes known as sieve tubes connected by sieve plates. Xylem is made up of rather tougher tubes called vessels and tracheids. The development of these conducting elements from a group of relatively undifferentiated cambial cells (Figure 2-17) involves the later stages

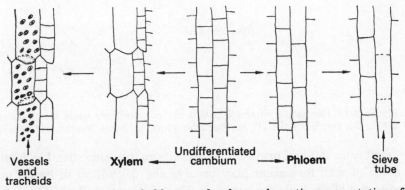

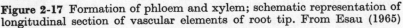

Vessels and tracheids   Xylem ⟵ Undifferentiated cambium ⟶ Phloem   Sieve tube

Figure 2-17 Formation of phloem and xylem; schematic representation of longitudinal section of vascular elements of root tip. From Esau (1965)

of wall formation; this is followed by gradual dissolution of the cytoplasmic contents to leave a series of conducting pores. The morphological development of xylem elements has been correlated with the gradual replacement of pectin by hemicellulose. The xylose:arabinose ratio of regions of sycamore stem, for example, rises some 40-fold during development. During further stages of xylem maturation, the wall thickens as deposition of lignin (which is absent from phloem) takes place. Phloem tissue, on the other hand, is characterized by the capacity to synthesize callose (a 1-3-β-linked glucose polymer), which appears to be laid down in the pores of the sieve plates.

As yet enzymic changes accompanying the formation of the

various polysaccharides have been little studied. It has recently proved possible to grow cambial-like callus tissue in culture. Under appropriate conditions, differentiation into phloem and xylem cells occurs. Such a system should prove useful for analysis of enzymic variation during development. It should also be possible to study the factors involved in phloem and xylem induction by this means. Already it appears that sucrose and two hormones, auxin (indol-3-acetic acid) and a kinin (Figure 2-14) are necessary. Their relative concentration in the culture medium selectively affects the course of phloem and xylem differentiation. After some time, the cultured cells themselves begin to synthesize auxin, so that a kind of positive feedback system is set up. The effect of hormones in stimulating developmental events is widespread. The action of gibberellin in germination has already been mentioned, and there are several other examples. Some aspects of hormonal induction of differentiation are discussed in Section 5-1.

By careful control of the composition of the culture medium, differentiated phloem cells of carrot root can be induced to 're-differentiate' and to develop into a whole carrot plant. This is described in Section 7-1(b).

(ii) *Pollen and Root Hair Cells*. Two examples may be cited of specialized cells in which the appearance of characteristic proteins has been followed. The cells have certain structural similarities in that each is situated at an extreme growing tip of plants.

The development of pollen has been studied in culture. Lily anthers contain microspores which give rise to pollen. Microspores have a life of several weeks, during which they undergo one cell division. This is preceded by the synthesis of thymidine kinase, an enzyme specifically involved in DNA synthesis (Figure 6-3). Appearance of thymidine kinase in cultured anthers is prevented by inhibitors of protein synthesis. Inhibitors of RNA synthesis also halt production of the enzyme, rather like the effect of actinomycin D on developing bacterial spores (Section 2-1(c)) and slime moulds (Section 2-2(b)).

Root hairs probably play an important part in the uptake of water and nutrients from soil by increasing the surface area of the root system. In animals the convolutions of the small intestine serve a similar purpose. The cells from which hairs are formed (trichoblasts) alternate with inactive cells in long rows in the epidermal layer of certain grasses. Nucleotidase (which hydrolyses AMP to adenosine) has been detected in trichoblasts, but is absent from adjacent cells (Figure 2-18). The enzyme may play a part in

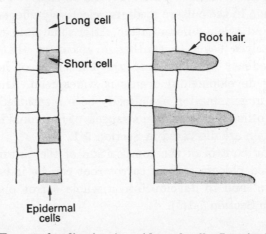

Figure 2-18 Enzyme localization in epidermal cells. Longitudinal section of epidermal layer of root. Shaded area shows histochemical localization of nucleotidase. This is restricted to short epidermal cells, which subsequently form hairs. From Avers (1961). *American J. Bot.*, **48**, 137 and Esau (1965)

the breakdown of nucleotides prior to absorption of phosphorus from the soil. In general, however, transport processes in root hairs have not yet been investigated to the extent that uptake of nutrients from the small intestine of animals has been studied (see Section 5-1(c)).

2-5 Concluding Remarks

Several examples of the emergence or alteration of proteins during cellular differentiation have been encountered. Biochemical analysis is hampered unless the systems can be studied under defined conditions. That is why bacterial sporulation, differentiation of slime mould myxamoebae and the germination of seeds, the differentiation of phloem and xylem and the development of pollen in culture, are systems most likely to yield answers to some of the outstanding questions. These may be summarized as (i) the mechanism underlying variation in protein synthesis, and (ii) the nature of the stimuli that cause such changes. Current views on these topics are discussed in Chapters 7 and 8.

The reader may conclude that, in general, there are fewer instances of cell type specificity in plants than will be seen to occur in animals (see Chapters 3–5). If it is true that plant cells do not differentiate to the extent that animal cells do, this might explain why it is easier to grow whole plants from single cells at terminal stages of development (see Section 7-1).

Summary

(a) Bacterial Spores

The transformation of a vegetative bacterial cell into a *spore*, called *sporulation*, is initiated by lack of nutrients. Spores are *resistant* to *environmental change*, such as humidity, temperature and ionizing radiation. Spores have a thickened wall, a low moisture content and a high concentration of dipicolinic acid and calcium ions. Some *enzymes*, and *proteins* in general, have *altered properties* compatible with heat and radiation resistance. The components typical of spores are synthesized in *sequence* after the cell has been committed to sporulation.

(b) Slime Moulds

Vegetative, unicellular myxamoebae of *D. discoideum* aggregate into a pseudoplasmodium, which *differentiates* into *spores* and *stalk cells* when nutrients are exhausted. *Aggregation* is *initiated* by a chemical substance, tentatively identified as *cyclic AMP*, which is only secreted by a fraction of the cells. Differentiation is accompanied by the *synthesis* of *specific enzymes*, such as UDP-glucose pyrophosphorylase, UDP-galactose transferase and trehalose-6-phosphate synthetase. They appear at *discrete times* after *commitment* to aggregation.

(c) *Plants*

Germination of *seeds* is accompanied by *synthesis of specific degradative enzymes* such as α-amylase. Synthesis is *initiated by* the secretion of *hormones*. The *properties* of several *enzymes* change during germination and further development.

The formation of plant *cell walls* is accompanied by a changing pattern of the constituent polysaccharides. Differentiated cells, such as phloem or xylem of conducting tissue or endodermal cells in roots, are characterized by walls of different chemical composition. *Specific enzymes* have been implicated to account for these changes. *Development* of phloem and xylem is dependent on the presence of certain *hormones*.

Maturation of pollen and root hair cells involves the synthesis of different *enzymes* such as thymidine kinase and nucleotidase respectively.

In general, *plant* cells appear to be *less differentiated* than *animal cells.*

Selected Bibliography

Bacterial Spores

General Reading

Gould, G. W. and A. Hurst (Ed.) (1969). *The Bacterial Spore.* Academic Press, London–New York

Keynan, A. (1969). The Outgrowing Bacterial Endospore. *Current Topics in Dev. Biol.*, **4**, 1

Murrell, W. G. (1967). The Biochemistry of the Bacterial Endospore. *Adv. Microb. Physiol.*, **1**, 133

Sussman, A. S. and H. O. Halvorson (1966). *Spores: their Dormancy and Germination.* Harper & Row, New York

Specific Aspects

Kornberg, A., J. A. Spudich, D. L. Nelson and M. P. Deutscher (1968). Origin of Proteins in Sporulation. *Ann. Rev. Biochem.*, **37**, 51

Mandelstam, J. (1969). Regulation of Bacterial Spore Formation. *Symp. Soc. Gen. Microbiol.*, **19**, 377

Warren, S. C. (1968). Sporulation in *Bacillus Subtilis.* Biochemical Changes. *Biochem. J.*, **109**, 811

Slime Moulds and Water Moulds

General Reading

Baldwin, H. H. and H. P. Rusch (1965). The Chemistry of Differentiation in Lower Organisms. *Ann. Rev. Biochem.*, **34**, 565

Bonner, J. T. (1967). *The Cellular Slime Molds*, Second Ed. Princeton University Press, New Jersey

Cantino, E. C. and J. S. Lovett (1964). Non-filamentous aquatic fungi: Model Systems for Biochemical Studies of Morphological Differentiation. *Adv. Morphogen.*, **3**, 33

Specific Aspects

Bonner, J. T. (1965). Evidence for the Formation of Cell Aggregates by Chemotaxis in the Development of the Slime Mold *Dictyostelium Discoideum*. In *Molecular and Cellular Aspects of Development* (Ed. E. Bell), Harper & Row, New York, p. 40

Bonner, J. T. (1969). Hormones in Social Amoebae and Mammals. *Scientific American*, **220**, No. 6, p. 78

Gerisch, G. (1968). Cell Aggregation and Differentiation in Dictyostelium. *Current Topics in Dev. Biol.*, **3**, 157

Gregg, J. H. (1965). Regulation in the Cellular Slime Molds. *Dev. Biol.*, **12**, 377

Miller, Z. I., J. Quance and J. M. Ashworth (1969). Biochemical and Cytological Heterogeneity of the Differentiating Cells of the Cellular Slime Mold *Dictyostelium discoideum*. *Biochem. J.*, **114**, 815

Sussman, M. (1967). Evidence for Temporal and Quantitative Control of Genetic Transcription during Slime Mold Development. *Fed. Proc.*, **26**, 77

Sussman, M. and R. Sussman (1969). Patterns of RNA Synthesis and of Enzyme Accumulation and Disappearance during Cellular Slime Mold Cytodifferentiation. *Symp. Soc. Gen. Microb.*, **19**, 403

Wright, B. E. (1965). Control of Carbohydrate Synthesis in the Slime Mold. In *Developmental and Metabolic Control Mechanisms and Neoplasia*. The Williams and Wilkins Company, Baltimore, Maryland, p. 296

Plants

General Reading

Bonner, J. (1965). Development. In *Plant Biochemistry* (Ed. J. Bonner and J. E. Varner), Academic Press, London–New York, p. 850

Clowes, F. A. L. and B. E. Juniper (1968). *Plant Cells*. Blackwell Scientific Publications, Oxford

Esau, K. (1965). *Plant Anatomy*, Second Ed. John Wiley & Sons, London–New York.

Fox, J. E. (1968). *Molecular Control of Plant Growth*. Prentice-Hall International, Englewood Cliffs, New Jersey

Greulach, V. A. and J. E. Adams (1967). Plants. *An Introduction to Modern Botany*. John Wiley & Sons, London–New York

Heslop-Harrison, J. (1967). Differentiation. *Ann. Rev. Plant Physiol.*, **18**, 325

Torrey, J. G. (1967). *Development in Flowering Plants*. The Macmillan Company, New York

Germination; Hormones

Filner, P. and J. E. Varner (1967). A Test for the *De Novo* Synthesis of Enzymes: Density-Labelling with H_2O^{18} of Barley α-Amylase Induced by Gibberellic Acid. *Proc. Nat. Acad. Sci.*, **58**, 1520

Johri, M. M. and J. E. Varner (1968). Enhancement of RNA Synthesis in Isolated Pea Nuclei by Gibberellic Acid. *Proc. Nat. Acad. Sci.*, **59**, 269

Key, J. L. (1969). Hormones and Nucleic Acid Metabolism. *Ann. Rev. Plant Physiol.*, **20**, 449

Marré, E. (1968). Ribosome and Enzyme Changes During Maturation and Germination of the Castor Bean Seed. *Current Topics in Dev. Biol.*, **2**, 76

Van Overbeek, J. (1968). The Control of Plant Growth. *Scientific American*, **219**, No. 1, p. 75

Varner, J. E. (1965). Seed Development and Germination. In *Plant Biochemistry* (Ed. J. Bonner and J. E. Varner), Academic Press, London–New York, p. 763

Chloroplasts

Criddle, R. S. (1969). Structural Proteins of Chloroplasts and Mitochondria. *Ann. Rev. Plant Physiol.*, **20**, 239

Hall, D. O. and Whatley, F. R. (1967). The Chloroplast. In *Enzyme Cytology* (Ed. D. B. Roodyn), Academic Press, London–New York, p. 181

Park, R. B. (1967). Control of Membrane Differentiation and Quantum Conversion Efficiency in Chloroplasts. In *Organisational Biosynthesis* (Ed. H. J. Vogel, J. O. Lampen and V. Bryson), Academic Press, London–New York, p. 373

Cell Walls

Mühlethaler, K. (1961). Plant Cell Walls. In *The Cell* (Ed. J. Brachet and A. E. Mirsky), Academic Press, London–New York, Vol. 2, p. 85

Mühlethaler, K. (1967). Ultrastructure and Formation of Plant Cell Walls. *Ann. Rev. Plant Physiol.*, **18**, 1

Northcote, D. H. (1969a). Fine Structure of Cytoplasm in Relation to Synthesis and Secretion in Plant Cell Walls. *Proc. Roy. Soc. B*, **173**, 21

Northcote, D. H. (1969b). The Synthesis and Metabolic Control of Polysaccharides and Lignin during the Differentiation of Plant Cells. In *Essays in Biochemistry* (Ed. P. N. Campbell and G. D. Greville), Academic Press, London–New York, Vol. 5, p. 89

Rogers, H. J. and H. R. Perkins (1968). *Cell Walls and Membranes*. E. & S. N. Spon, London

Specialized Cells

Fleet, D. S. van (1961). Histochemistry and Function of the Endodermis. *Bot. Rev.*, **27**, 165

Jacobs, W. P. (1970). Regeneration and Differentiation of Sieve-Tube Elements. *Intern. Rev. Cytol.*, **28**, 239.

Northcote, D. H. (1969c). Growth and Differentiation of Plant Cells in Culture. *Symp. Soc. Gen. Microb.*, **19**, 333

Shannon, L. M. (1968). Plant Isoenzymes. *Ann. Rev. Plant Physiol.*, **19**, 187

Stern, H. and Y. Hotta (1968). Biochemical Studies of Male Gametogenesis in Liliaceous Plants. *Current Topics in Dev. Biol.*, **3**, 37

CHAPTER 3

Differentiation in animals: sea urchins and frogs

The last chapter contained few examples in which differentiated structure and function were directly explicable in terms of specific proteins; the localization of particular enzymes in certain cells lacked significance in that the physiological role was often not apparent. Although there are even more descriptions of such enzymic variations in animals, there are also a number of cases in which development of unique structure and function can be correlated with alteration in specific enzymes and other proteins. It is with such examples that the next two chapters will be concerned.

3-1 Embryology

The anatomical and histological development of a fertilized animal egg into a mature embryo, larva or foetus, which is perhaps the most striking example of differentiation, has been studied for many years. It was soon found that the early stages in human embryos resembled those in other mammals. This is perhaps not surprising since human egg cells (like individual adult cells) are hardly distinguishable from those of a mouse or a rabbit, each, incidentally, being several hundred-fold larger than adult cells. Great interest in embryology ensued and it was thought that here was a key, not only to the emergence of human form and function, but to the development of man from other species. One of the most interesting facts to emerge was the discovery that fertilized eggs, after a sufficient number of divisions, come to form three histologically distinct areas called endoderm, mesoderm and ectoderm (Figure 3-1). Each of these layers contributes to specific tissues in

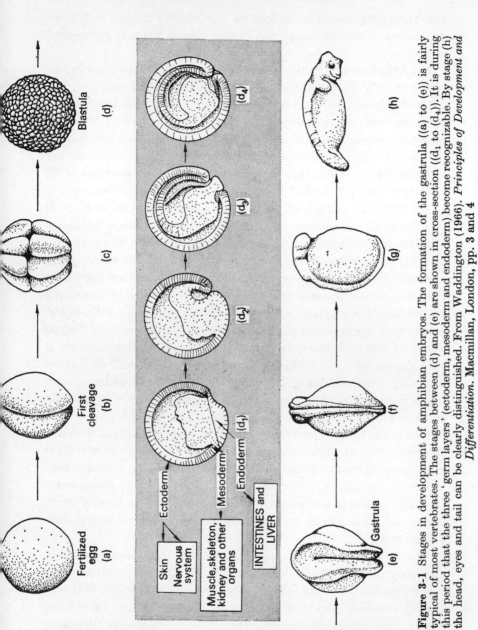

Figure 3-1 Stages in development of amphibian embryos. The formation of the gastrula ((a) to (e)) is fairly typical of most vertebrates. The stages between (d) and (e) are shown in cross-section ((d_1) to (d_4)). It is during this period that the three 'germ layers' (ectoderm, mesoderm and endoderm) become recognizable. By stage (h) the head, eyes and tail can be clearly distinguished. From Waddington (1966). *Principles of Development and Differentiation*. Macmillan, London, pp. 3 and 4

the adult; for example, endoderm contributes to gut and liver, mesoderm to muscle, bones and blood, and ectoderm to nerve cord and skin.

Much descriptive detail concerning the development of embryos, based mainly on histological techniques, has since emerged, but little is relevant to the present discussion since few biochemical analyses were carried out. This has not prevented speculation about the biochemistry of differentiation. The simplest hypothesis, which one might take as the origin of current views concerning differentiation, is that the fertilized egg contains no enzymes at all but that these are formed in sequence as the embryo develops.

By 1931 this was felt to be unacceptable, since even an egg cell was found to have some enzymes, such as a protease, a lipase, an amylase and some oxidizing enzymes. Nevertheless, the idea of a gradual emergence of new enzymes was popular for many years and still has merit. It may be contrasted with another view that sees differentiation as a *loss* of function, in this case genotypic rather than phenotypic, as specialized tissues develop. Although it is true that in cases such as red cell formation (Section 4-3(*a*)) some enzymes and indeed whole structures are lost, the mechanism is now known not to be by loss of specific genes (Section 7-1). Rather one may view development, including that of the embryo, as a changing pattern of enzyme activities brought about by specific activation or inhibition of existing genes (Section 8-1).

One of the reasons for the dearth of biochemical data is that it is difficult to get enough material for study, the embryo of a typical experimental animal such as a rat or a mouse being rather small. However, by pooling embryos from groups of rats at presumedly identical stages of pregnancy, this drawback can be largely overcome and results from such studies are beginning to accumulate (Chapter 4).

The fact that amphibian and many other eggs develop outside the body and are available in very large numbers, is an obvious advantage and has led to their rather wide use as tools in embryological research. Two groups, sea urchins and frogs, have been particularly extensively investigated. Most of the results that have been obtained concern DNA and RNA synthesis rather than protein synthesis. However, in so far as synthesis of any molecule implies the functioning of specific enzymes, and because nucleic acid synthesis underlies the very mechanism of protein synthesis, discussion of nucleic acid metabolism is included in this chapter. Sea urchin eggs have one particular advantage over amphibian eggs in that they are much more permeable to added macromolecular precursors such as amino acids and nucleosides.

3-2 Embryological Development of Sea Urchins

(a) Cleavage

One of the first morphological events following fertilization of an egg cell is *cleavage* (Figure 3-2). In the case of the American sea

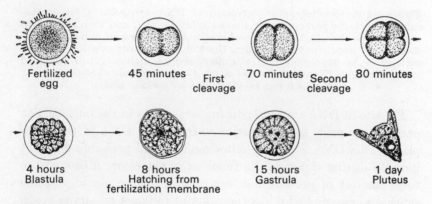

Fertilized egg   45 minutes   First cleavage   70 minutes   Second cleavage   80 minutes

4 hours Blastula   8 hours Hatching from fertilization membrane   15 hours Gastrula   1 day Pluteus

Figure 3-2 Development of sea urchin embryos. The approximate time taken to reach the various stages is given. Note the similarity of blastula formation to that in a vertebrate (see Figure 3-1). From Harvey (1956)

urchin *Arbacia punctulata* the fertilized egg (approximately 70 μm in diameter) cleaves some ten times to produce about a thousand cells without overall change in its size. At this stage, seven to eight hours after fertilization, the blastula is formed. The total amount of cell components such as protein, fat and carbohydrate does not change during this period but that of DNA increases in proportion to the number of new cells formed. The synthesis of DNA has been studied spectrophotometrically and by autoradiographic detection of incorporation of a radioactive precursor, such as thymidine, at various times after fertilization (Figure 3-3).

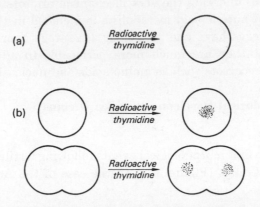

Figure 3-3 Autoradiographic detection of DNA synthesis. Cells are incubated with radioactive thymidine for a brief period and then applied to a microscope slide. After fixing and washing, incorporated radioactivity is detected by exposure to x-ray film. Black dots mark the position of radioactivity in the developed film. (a) unfertilized sea urchin egg; (b) fertilized sea urchin eggs. Note that the radioactivity appears only over the nuclear area, which can be visualized by specific stains

The site of DNA synthesis during cleavage is in the nucleus. Sea urchin eggs prior to fertilization also possess DNA in the cytoplasm. This DNA is mainly mitochondrial and probably does not increase during cleavage; its function is not clear. Although the total amount of protein does not change during cleavage, some synthesis, measured by incorporation of radioactive amino acids, does take place after fertilization. However minor in amount relative to total protein, it could be of considerable significance and the nature of the protein is being investigated. Another constituent that one may expect to increase during cleavage is the plasma membrane, but few studies on this appear to have been carried out.

The energy for the processes associated with cell division is provided by oxidative metabolism and it is perhaps not surprising that a sharp increase in oxygen uptake follows fertilization. As long ago as 1910, Otto Warburg suggested that this is the 'sparking reaction' for further events. In the case of sea urchins this may be so, but it cannot have the general significance proposed by Warburg since some eggs, such as those of teleost fish, show no increased oxygen uptake on fertilization, while others, such as those of the annelid *Chaetopterus*, actually show a decline (Figure 3-4).

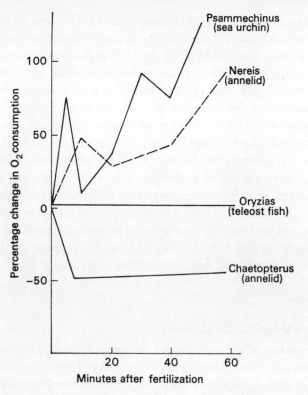

Figure 3-4 Changes in oxygen consumption of sea urchin and other eggs following fertilization. From Monroy (1967)

The change in respiration by sea urchins is not due to an increased synthesis of enzymes, which are present in quite large amounts before fertilization, but to activation of existing enzymes. The stimulus for this may be the ratio of ATP to ADP, which falls when demands for chemical energy occur. Another reason may be the movement of a particular enzyme, glucose-6-phosphate dehydrogenase, from an inactive particle-bound state into the soluble phase of the cytoplasm where it is active; or again the

formation of glucose-6-phosphate from glycogen, which is the form
in which carbohydrate synthesized during egg formation is stored,
may control the subsequent oxidation.

(b) Gastrulation

During the formation of the gastrula (12–15 hours after fertilization) the outer membrane which surrounds the blastula, known as
the fertilization membrane, is shed. This process is termed hatching. An enzyme which appears to be responsible for hatching has
been isolated and even crystallized; it is a protease and its appearance just before hatching has been shown to be due to synthesis of
new molecules. The mechanism of its specificity in degrading only
the protein of the fertilization membrane remains to be elucidated.

After the gastrula has formed, an increase in overall size accompanies the formation of a skeleton and culminates, one day after
fertilization, in a free swimming, well differentiated larva called a
pluteus. This then undergoes metamorphosis and eventually forms
a young sea urchin. Few biochemical investigations of these events
have been carried out. In amphibia such as frogs, on the other
hand, metamorphosis has been rather extensively studied.

3-3 Frogs

(a) Embryological Development

The embryonic development of frogs is morphologically and
biochemically similar to that of sea urchins. Certain processes such
as RNA metabolism have been particularly investigated. Most of
the examples which follow are taken from *Xenopus laevis*, the
South African clawed toad.

As in the sea urchin, there is extensive DNA synthesis during
cleavage. The period of the cell cycle during which DNA synthesis
occurs (generally called the synthetic or S period) has been
analysed. When cell division is rapid (fifteen minutes), as in
cleavage, the S period is short (twelve minutes), but as the rate of
cell division falls, so the duration of S increases; at the neurula
stage, for example, it lasts six hours in cells which divide every
twenty-four hours. Yet the amount of DNA being copied is the

same. The length of S may therefore depend on other processes, such as transcription of DNA to RNA.

During cleavage, certain areas of the embryo, such as the endoderm at one end, divide more slowly than others and DNA synthesis is correspondingly less frequent. This is one of the earliest biochemical signs of differentiation.

The onset of DNA synthesis is due to the activation of existing enzymes rather than to the synthesis of new enzymes, since it takes place in the presence of inhibitors which block the synthesis of proteins and RNA. The same is true of DNA synthesis in sea urchins. The nature of the activating stimulus is unknown; it appears to coincide with the movement of certain cytoplasmic proteins into the nucleus (Section 8-2). While DNA synthesis is occurring during cleavage, little if any RNA is made; in contrast, much RNA but no DNA is synthesized during the prior formation of the female unfertilized egg cell (oogenesis). In late cleavage, as the frequency of DNA synthesis declines, RNA production begins once more.

It is possible to distinguish the three main types of RNA (Figure 8-1). These are transfer or tRNA (the species which transports activated amino acids from cytoplasm to ribosomes during protein synthesis), ribosomal or rRNA (which together with proteins constitutes ribosomes and accounts for about 80 per cent of the total cellular RNA) and DNA-like or dRNA (which on average reflects the base composition of the total DNA and includes messenger or mRNA). Conventional separation methods, such as centrifugation through density gradients of sucrose, electrophoresis in acrylamide gel and chromatography on columns of cellulose or kieselgühr, have been used to analyse RNA from eggs which have been incubated with radioactive precursor such as uridine or phosphate.

Such studies have shown that when RNA synthesis begins in late cleavage, dRNA is first formed, followed quite soon by rapid tRNA synthesis. rRNA synthesis does not begin until the gastrula stage (Figure 3-5). Even then rRNA synthesis cannot be of paramount importance to the embryo since whole tadpoles have been obtained from mutants of *Xenopus laevis* which are unable to produce any rRNA at all. In other words, the rRNA made during

oogenesis suffices up to this stage. The synthesis of rRNA is
separated from that of tRNA and mRNA not only in time but also
spatially, since it takes place in the morphologically distinct
region of the nucleus termed the nucleolus. Sequential synthesis
of specific RNA molecules has been observed in sea urchins also.

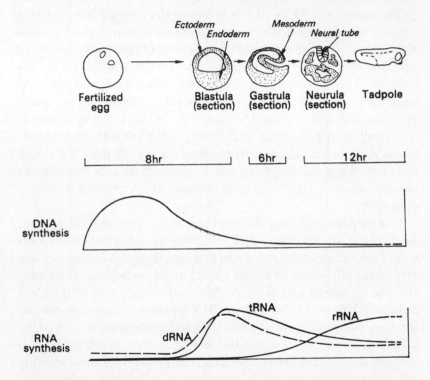

Figure 3-5 Nucleic acid synthesis in amphibian embryos. Different classes
of RNA are synthesized at specific stages in development. From Gurdon
(1968)

Since RNA polymerase is believed to be responsible for the
synthesis of all types of RNA, the selective synthesis of one or
other type must reflect activation of the corresponding genes.
Certain specific protein cofactors associated with RNA polymerase may be involved in such regulation. Selective gene
activation is discussed more fully in Section 8-1(*b*).

(b) *Metamorphosis*

A key stage in the growth of frogs is their development from tadpoles, a change known as metamorphosis (Figure 3-6). A

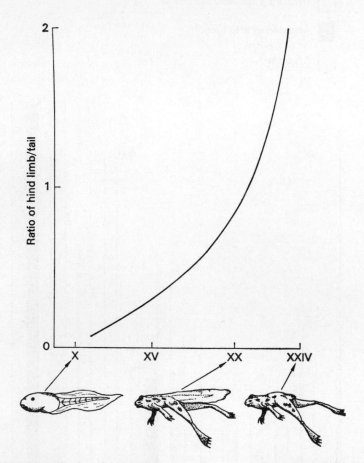

Figure 3-6 Ratio of hind limb to tail as an index of amphibian metamorphosis (*Rana catesbeiana*). From Brown and Cohen (1958)

characteristic feature of tadpoles which is absent in frogs is their tail. Tail resorption involves the degradation of its collagen, a protein typical of connective tissue (Section 4-2(a)), and one of the earliest proteins to be synthesized in developing tadpoles. It is

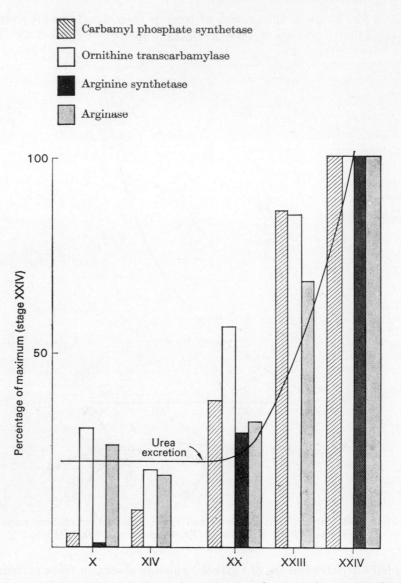

Figure 3-7 Excretion of urea and appearance of urea cycle enzymes (see Figure 3-8) during amphibian metamorphosis (*R. catesbeiana*). All values are expressed relative to that attained at stage XXIV (see Figure 3-6). From Brown and Cohen (1958)

therefore interesting that certain proteolytic enzymes including a collagenase, appear during metamorphosis. As in the hatching enzyme of sea urchins, the specificity of these enzymes has not been clarified. The *disappearance* of certain cells or their contents is as typical a developmental event as the *appearance* of specific cell types. The phenomenon of ageing, the biochemistry of which is as yet poorly understood, probably involves a similar loss of certain functions in particular cells, though the developmental stage at which ageing occurs is generally unpredictable.

Another difference is that tadpoles are aquatic animals whereas frogs are essentially terrestrial. This means that tadpoles like fish can excrete their waste products continuously into the environment by diffusion through their gills. The major product of nitrogen metabolism is ammonia, which is toxic to all animal cells. This does not matter to tadpoles and fish since the ammonia is removed as fast as it is formed. In frogs, as in other terrestrial animals, toxic levels would soon accumulate. Ammonia is detoxified in one of two ways. It is either converted into uric acid which is excreted as a solid, or into urea which is excreted in aqueous solution. Birds and reptiles which are short of water excrete uric acid, whereas most other terrestrial animals excrete urea. The synthesis of either uric acid or urea requires chemical energy and the process is therefore avoided where possible.

Urea synthesis, which occurs almost exclusively in the mitochondria of the liver, has been studied in metamorphosing tadpoles of *Rana catesbeiana*. It is seen that urea production rises very sharply at about stage XX (Figure 3-7) when the ratio of length of hind limb to length of tail equals one. The synthesis of urea from ammonia involves several enzymes (Figure 3-8) and these have been found to increase just before the onset of urea production (Figure 3-7). The increase is due to the production of new enzyme and, unlike the enzyme catalysing polysaccharide synthesis in differentiating slime moulds, for example (Section 2-2(b)), synthesis continues for some time at a lower rate while the size of the liver increases. The production of these enzymes is triggered by the appearance of the hormone thyroxine which is more fully discussed in Section 5-1. Many other enzymes also change in activity or in

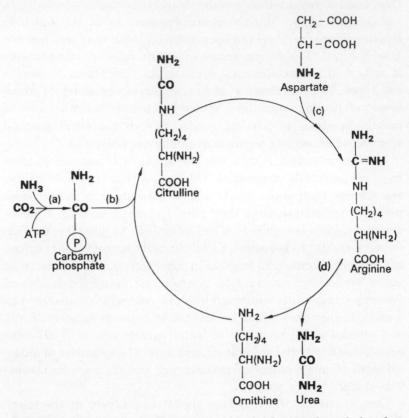

Figure 3-8 The urea cycle. The enzymes involved are carbamyl phosphate synthetase (a), ornithine transcarbamylase (b), arginine synthetase (c) and arginase (d)

amount during metamorphosis but their physiological significance is not always so clear.

3-4 Concluding Remarks

It has been seen that cleavage is accompanied by the sequential functioning of specific enzymes such as DNA and RNA polymerase. Many other enzymes such as the hatching protease or oxidizing enzymes concerned with carbohydrate utilization also appear at specific times during the development of the embryo.

In addition to this *temporal* control of biochemical activity, there is a *regional* or *spatial* control which is responsible for the emergence of specialized structures at different sites within the embryo. Although it is known that commitment to regional specialization occurs very early on, probably quite soon after fertilization or even before, the mechanism is unknown. Gradients in the internal concentration of molecules which trigger off particular enzymes may be part of the answer (Section 8-2(*b*)). Clearly this is an important aspect of differentiation which awaits attack by the biochemist.

3-5 Summary

(a) *Embryological Development*

Embryological studies of sea urchins and frogs have shown that when eggs undergo *cleavage* following fertilization, *DNA* is *synthesized*. There is little net synthesis of protein, carbohydrate, fat or RNA which are laid down prior to fertilization. Some time after the commencement of cleavage, *RNA synthesis* starts; it increases as DNA synthesis begins to decline. In frogs *dRNA* is formed *first*, followed by *tRNA* and then by *rRNA*. The enzymes for synthesizing DNA and RNA are probably present in the egg prior to fertilization; how they are prevented from acting is not yet known.

During the gastrulation of sea urchin embryos, the blastula membrane is ruptured; this is due to the *synthesis* of a specific *protease*.

(b) *Metamorphosis*

The *formation* of *specific enzymes* has been correlated with certain developmental events during the metamorphosis of tadpole into frog. A *collagenase* has been implicated in the resorption of tadpole tail, while the *enzymes* catalysing *urea formation* are *synthesized* just before the animal changes from excreting its nitrogen as ammonia to excreting it as urea.

Selected Bibliography

General Reading

Balinski, B. I. (1965). *An Introduction to Embryology.* W. B. Saunders Company, Philadelphia

Davidson, E. H. (1969). *Gene Activity in Early Development*. Academic Press, London–New York

Deuchar, E. M. (1966). *Biochemical Aspects of Amphibian Development*. Methuen, London

Fischberg, M. and A. W. Blackler (1961). How Cells Specialise. *Scientific American*, **205**, No. 3, p. 124

Needham, J. (1931). *Chemical Embryology*. Oxford University Press, Oxford

Spemann, H. (1938). *Embryonic Development and Induction*. Yale University Press, New Haven, Conn.

Early Embryonic Development

Brachet, J. (1967a). Behaviour of Nucleic Acids During Early Development. In *Comprehensive Biochemistry* (Ed. M. Florkin and F. M. Stotz), Vol. 28, p. 23, Elsevier Publishing Co., Amsterdam

Brachet, J. (1967b). Biochemical Changes During Fertilization and Early Development. In *CIBA Foundation Symposium* (Ed. A. V. S. de Reuck and J. Knight), J. & A. Churchill, London, p. 39

Denis, H. (1968). Role of Messenger Ribonucleic Acid in Embryonic Development. *Adv. Morphogen.*, **7**, 115

Gross, P. R. (1967). RNA Metabolism in Embryogenesis. In *Macromolecular Synthesis and Growth* (Ed. R. A. Malt), J. &. A. Churchill, London, p. 185

Gurdon, J. B. (1968). Nucleic Acid Synthesis in Embryos and its Bearing on Cell Differentiation. In *Essays in Biochemistry* (Ed. P. N. Campbell and G. D. Greville), Vol. 4, p. 25

Harvey, E. B. (1956). *The American Arbacia and Other Sea Urchins*. Princeton University Press, New Jersey

Monroy, A. (1967). Fertilisation. In *Comprehensive Biochemistry* (Ed. M. Florkin and E. H. Stotz), Elsevier Publishing Co., Amsterdam, Vol. 28, p. 1

Moog, F. (1967). Enzyme Development in Relation to Functional Differentiation. In *Biochemistry of Animal Development* (Ed. R. Weber), Academic Press, London–New York, Vol. 1, p. 307

Nemer, M. (1967). Transfer of Genetic Information during Embryogenesis. *Progress in Nucleic Acid Res.*, **7**, 243

Sherbet, G. V. and M. S. Lakshmi (1967). Structural Organization and Embryonic Differentiation. *Intern. Rev. Cytol.*, **22**, 147

Tiedemann, H. (1966). The Molecular Basis of Differentiation in Early Development of Amphibian Embryos. *Current Topics in Dev. Biol.*, **1**, 85

Yamada, T. (1961). A Chemical Approach to the Problem of the Organiser. *Adv. Morphogen.*, **1**, 1

Metamorphosis

Brown, G. W. and P. P. Cohen (1958). Biosynthesis of Urea in Metamorphosing Tadpoles. In *Chemical Basis of Development* (Ed. W. D. McElroy and B. Glass), Johns Hopkins Press, Baltimore, Md., p. 495

Weber, R. (1967a). Biochemistry of Amphibian Metamorphosis. In *Comprehensive Biochemistry* (Ed. M. Florkin and E. M. Stotz), Elsevier Publishing Co., Amsterdam, Vol. 28, p. 145

Weber, R. (1967b). Biochemistry of Amphibian Metamorphosis. In *Biochemistry of Animal Development* (Ed. R. Weber), Academic Press, London–New York, Vol. 2, p. 227

Useful Films on Embryology

Beginnings of vertebrate life (1963). Encyclopaedia Britannica Films (U.S.A.)

Frog development—fertilization to hatching (1966). Educational Services (U.S.A.)

The development of a fish embryo (1962). National Film Board of Canada

Development of amphibian embryo. Institute of Animal Genetics, Edinburgh (U.K.)

CHAPTER 4

Differentiation in animals: birds and mammals

The early embryology (formation and fertilization of eggs, cleavage, formation of blastula and gastrula) of higher animals is very similar to that of sea urchins and frogs and there is little reason to doubt that the major biochemical events are not the same also. That is to say that protein, carbohydrate and fat are laid down during oogenesis. Following fertilization, which takes place within the animal, the reserves begin to be utilized. In the case of birds this lasts the entire period of embryological growth within the egg. In the case of mammals, food (mainly glucose and amino acids) is taken from the maternal circulation as soon as the connecting link or placenta is formed. During cleavage, DNA synthesis followed by RNA synthesis are the main biochemical events. From gastrulation onwards, as differentiation of structure and function become intense, synthesis of specific proteins accompanies the continuing synthesis of DNA and RNA.

In this chapter some biochemical changes which occur at birth and soon after will be examined (Section 4-1). The type of bird which has been most widely studied is the chicken; the rat has proved to be a useful experimental animal among mammals. In certain cases, it has been possible to study the later stages of embryological development in the test tube instead of the egg, and some of the findings regarding cartilage, kidney and pancreas formation will be discussed. Differences in the proteins of completed adult organs such as liver, heart and muscle, which are found in all higher animals, will be considered next (Section 4-2) and will lead to a discussion of isoenzymes. Finally some aspects of adult development, that is to say the continued formation of differentiated cells such as the red and white blood cells, will be described (Section 4-3); this will include a consideration of antigens and antibodies.

4-1 Early Development

(a) Birds

One of the first proteins to become detectable that has cell-type specificity is haemoglobin. It has been observed only thirty-six hours after the commencement of embryonic development. Another protein that appears as early as this is myosin, which with actin constitutes the contractile muscle protein actomyosin. By careful dissection of embryos, different areas can be tested for the presence of myosin. Heart muscle myosin is first found over most of the embryo, but as development proceeds, it becomes increasingly localized in the area destined to become the embryonic heart.

Generally speaking, most enzymes increase in activity during the later stages of embryonic growth. Thus proteases and peptidases increase many-fold between five and ten days after hatching, some oxidizing enzymes such as cytochrome oxidase and succinic dehydrogenase rise steadily up to twenty days, while others such as lactic, malic and glutamic dehydrogenase increase up to fifteen days but then decline. At the same time the localization of glutamic dehydrogenase shifts from cytoplasm to mitochondria (compare a reverse shift of glucose-6-phosphate dehydrogenase in the sea urchin). Some enzymes like aldolase remain constant; others alter in specificity, alkaline phosphatase of the diverticulum being replaced by acid phosphatase (note the phenomenon of isoenzymes discussed in Section 4-2(b)). Entire metabolic sequences may alter; for example, the pattern of nitrogen excretion changes from the 'primitive' state of ammonia excretion to that of urea excretion, followed finally by the excretion of uric acid which continues throughout adult life. The reason for this is that in an enclosed egg, even urea becomes too toxic if it is not removed, whereas uric acid can be deposited as insoluble crystals without affecting the metabolism of the egg.

Like most eggs, chick eggs are low in carbohydrate compared to fat and protein, and embryonic chick liver uses fat as the main energy source. It makes extra glucose, which may be required for transport to the brain and for the synthesis of specific polysaccharides in other tissues, from amino acids and other non-carbohydrate precursors. This process, called gluconeogenesis, is in essence the reversal of glycolysis (lactate production from glucose) except at certain points where, for thermodynamic reasons, glycolytic and gluconeogenic reactions differ (Figure 4-1). One such point is the interconversion of fructose-6-phosphate and fructose-1, 6-diphosphate. The activity of fructose-1,6-diphosphatase (FDPase) is considered to be an index of gluconeogenesis and the enzyme is indeed twice as high in the embryo as in the young chick, which begins to feed on carbohydrate as soon as it is hatched. Conversely, phosphofructokinase (PFK) is ten times as high in a week-old chick as in the embryo. Unlike the phosphatase mentioned above (or the isoenzymes to be discussed later), the FDPase of adults has the same properties (pH optimum, magnesium activation and inhibition by AMP or excess FDP) as the embryonic enzyme and the decreased activity is therefore most likely due to reduced enzyme synthesis or increased degradation rather than to specific inhibition of existing enzyme.

(b) Rats

Unlike the developing chick embryo, the rat embryo or foetus has a plentiful supply of glucose through the placenta, which is a layer of specialized tissue linking the foetus to the maternal blood supply. The placenta maintains a constant supply of glucose by synthesizing some itself when maternal blood sugar is low. Other small molecules such as water, amino acid and vitamins pass the placenta, but fat droplets and bulky molecules such as most proteins do not. Hence in the rat foetus the main source of energy is carbohydrate and any necessary fat is made from glucose; in other words, the situation is the opposite of that in the embryonic chick.

At birth the suckling rat turns to a diet of milk, which consists of much fat but relatively little carbohydrate. One may therefore expect the enzymes oxidizing carbohydrate or converting it to fat to decrease as the rat is born. Two such enzymes, pyruvate kinase and ATP citrate lyase (Figure 4-1), do just that (Figure 4-2). Note

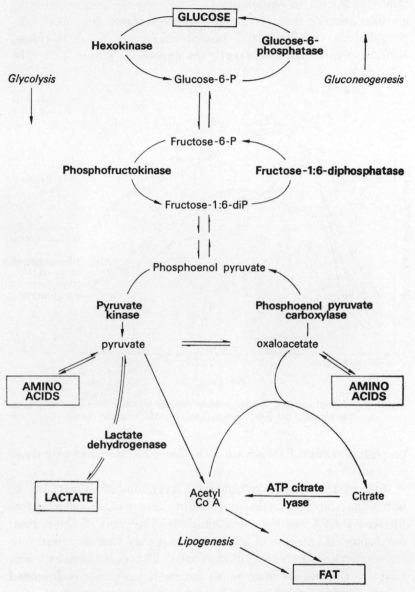

Figure 4-1 Enzymes of glycolysis and gluconeogenesis

that at about twenty days both activities rise rapidly again; this coincides with the end of lactation and the switch to grain and other foods rich in carbohydrate. The enzymes concerned with gluconeogenesis, such as glucose-6-phosphatase, fructose-1,6-diphosphatase (FDPase) and phosphoenolpyruvate carboxylase, alter in reverse manner exactly as expected (Figure 4-2). The

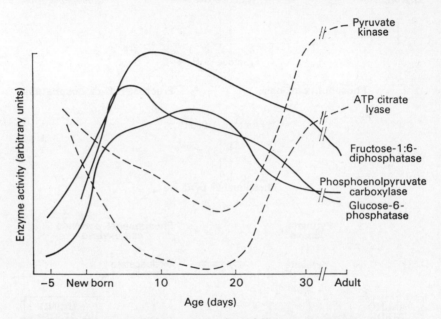

Figure 4-2 Enzymes of gluconeogenesis and lipogenesis (see Figure 4-1) in developing rat liver. From Vernon and Walker (1968)

properties of adult FDPase are, as in the chick, identical with those of foetal liver.

Most of the changes which have been described are due to nutritional circumstances and might be regarded as adaptive (Section 1-1(b)). On the other hand they are part of the normal developmental process of an animal, and may therefore justly be included in a discussion of differentiation. The mechanism by which adaptive changes are maintained during development is discussed below (Section 4-2(b)).

(c) Induction in Culture

Somite or cartilage and muscle-forming cells can be distinguished in a three-day-old chick. When such cells are dissected out and cultured under sterile conditions they form cartilage (chondrogenesis) within four days, provided they are in contact with embryonic spinal cord or notochord (one of the sites in the animal where cartilage formation subsequently occurs). Only somite cells respond and only spinal cord or notochord induce chondrogenesis. The *induction* (Section 1-1(*d*)) appears to be mediated by the release of a chemical substance, since somites and inducing factor can be separated by a filter and still be effective. Moreover extracts of spinal cord or notochord are active. The essential ingredient has not been identified but it may be a nucleotide, possibly linked to one of the sugars present in chondroitin sulphate which is a characteristic component of cartilage; the order of appearance

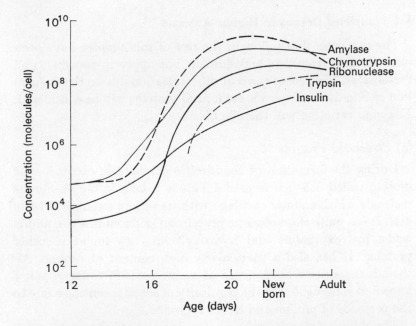

Figure 4-3 Formation of pancreatic proteins in developing mouse embryo. Note that specific proteins increase at different stages of development. From Wessels and Rutter (1969)

during development of the enzymes necessary for chondroitin sulphate formation remains to be elucidated.

Another system that has been studied in culture is the formation of embryonic mouse kidney tubules. Various embryonic tissues including spinal cord are effective inducers, but in this case material of higher molecular weight appears to be involved. This is also true of mouse pancreas, in which an embryonic protein fraction is necessary for the development of pancreatic form and function (as measured by the appearance of α-amylase, one of the enzymes characteristic of pancreatic development in the intact embryo (Figure 4-3)). Induction of several other morphogenetic events has been examined (see also Chapter 5) but in no case has a single substance been identified as the causative agent in the way that cyclic AMP has been shown to initiate the aggregation of slime mould myxamoebae (Section 2-2(c)).

4-2 Completed Organs in Higher Animals

The examples that follow in the rest of this chapter have been taken mainly from rats and humans but apply in essence to all vertebrates. First some cases of differentiation due to the acquisition or alteration of specific structural proteins will be considered. Enzymic variation will then be discussed.

(a) Structural Proteins

During the formation of chondroitin sulphate in cartilage, the protein called collagen is made. This is a long insoluble fibrous molecule which endows cartilage with its tensile strength (Figure 4-4). It is a quite characteristic protein since it contains two amino acids, hydroxyproline and hydroxylysine, not found in other proteins. It has also a particularly high content of proline. Although the structural sequence of amino acids in collagen is not known, it is thought that the regularity of shape is somehow due to the presence of proline and hydroxyproline.

The ability of muscle to contract and relax has been explained in terms of two specific proteins which it contains called actin and myosin. Although the amino acid composition of actin and myosin is not particularly striking, these proteins, like collagen, exist mainly as insoluble fibres. Electron microscopy and x-ray diffraction of muscle have shown that actin and myosin are attached and

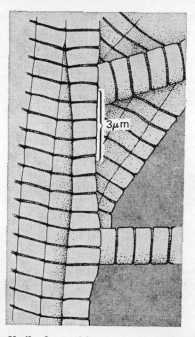

Figure 4-4 Collagen fibrils from skin, as seen under the electron microscope. The length of individual molecules, which are stacked side by side in rows and at right angles, is about 3 μm. From Gross (1961)

lie side by side in filaments which slide over each other as the muscle fibres lengthen and shorten (Figure 4-5).

There are three types of muscle fibre which are differentiated in terms of structure and function. Smooth or involuntary muscle is typical of the intestine, the movements of which are not regulated by conscious stimuli from the brain. In contrast, striated or voluntary muscle is responsible for all the conscious movement of arms, legs and other limbs. It is also referred to as skeletal muscle. 'Smooth' and 'striated' refer to the appearance of the muscle fibres under the microscope. The third type is heart muscle which is involuntary yet striated. Its structure is not the same as skeletal muscle, which is perhaps not surprising since its function is also different, being continuous at a regular rate throughout the life of the animal, whereas most skeletal muscle functions only sporadically. Although all types of muscle are controlled in rate of contraction by nervous impulses, conscious or unconscious, contraction

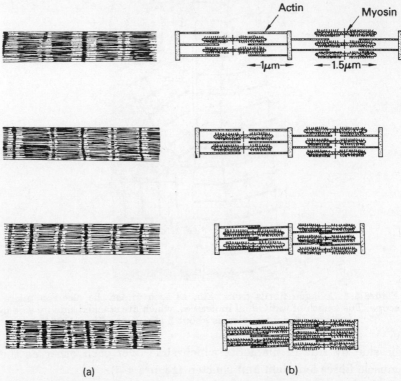

(a)   (b)

Figure 4-5 Muscle fibrils (see also Plate 1). During contraction the characteristic spacings seen under the electron microscope (a) shorten due to a sliding of the constituent actin and myosin filaments relative to one another (b). From Huxley (1965)

itself is a property solely of the actin and myosin filaments. Thus an embryonic heart begins to beat well before nerve fibres have become attached and it is possible to demonstrate contraction and relaxation of actomyosin removed from muscle and placed in solutions of particular salts.

One might expect to find certain differences in the actomyosin of the different types of muscle and this is actually so. The myosin of heart and skeletal muscle have been most clearly studied. They differ in solubility, molecular weight, in the firmness of their binding to actin and in their antigenicity. This last named property is most important and will be discussed later (Section 4-3(b)). The specific appearance of heart muscle myosin during embryonic development has already been referred to.

There are several other proteins which are not enzymes (but see Chapter 1) and which are specific to certain tissues. One can mention only a few, such as thyroglobulin, which is an iodinated protein of the thyroid and the precursor of the hormone thyroxine, albumin and globulins from blood serum, haemoglobin from red blood cells and many hormones that are proteins. Some of the blood proteins will be considered later in this chapter (Section 4-3); hormones are discussed in Chapter 5.

(b) Enzymes

(i) *Cell-specific Enzymes*. Reference has already been made to the localization of the enzymes catalysing urea formation in the liver of frogs. The same is true of birds and mammals. Several other metabolic reactions are confined to the liver. The formation of 'ketone bodies' (acetoacetate, acetone and β-hydroxybutyrate) is an example. Acetoacetate, which gives rise to acetone and β-hydroxybutyrate by decarboxylation and reduction respectively is formed by the pathway shown in Figure 4-6. The first two enzymes are present in other tissues such as the adrenals, since β-methyl-β-hydroxyglutaryl-CoA is the starting point for the synthesis of steroids. But only liver appears to contain the enzyme forming acetoacetate. Normally ketone bodies are utilized by muscle and other tissues and do not accumulate. However in starvation, or the disease diabetes mellitus, utilization is outstripped by an increased production of acetoacetate from the liver and ketone bodies appear in blood and urine, a condition known as ketosis. The ketone bodies are toxic to the body by virtue of their acidity. In general, however, one of the main functions of the liver is the reverse, that of detoxifying foreign substances that have found their way into the blood stream. For example, liver has dehydrogenases for oxidizing ethyl alcohol to acetaldehyde and to acetic acid or for turning D-amino acids into the corresponding keto acids, methylases for converting pyridine (C_5H_5N) into methyl pyridine ($C_5H_5NCH_3{}^+$) and a series of special hydroxylating enzymes for metabolizing such chemicals as aniline ($C_6H_5NH_2$)

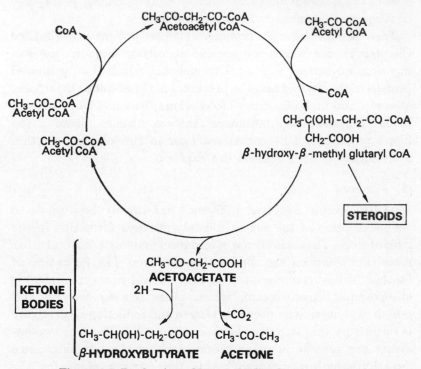

Figure 4-6 Production of ketone bodies from acetyl-CoA

into o- and p-amino phenols ($HOC_6H_4NH_2$). There is considerable species specificity. Thus phenylacetic acid is detoxified by conjugation with glutamine in man and the chimpanzee, with ornithine in the hen and with glycine or glucuronic acid in most other animals.

The liver is the main organ responsible for gluconeogenesis (see Section 4-1); kidney and placenta have a lesser capacity. Most of the enzymes involved in gluconeogenesis are present in other tissues like muscle where they catalyse the reverse process of glycolysis; but key enzymes such as fructose-1,6-diphosphatase (FDPase) and glucose-6-phosphatase (see Figure 4-1) are restricted to liver, kidney and the placenta.

The functional difference between organs is not always an all-or-nothing situation as in gluconeogenesis or the production of urea or ketone bodies. More often the pathways are the same but the rate at which they operate varies. Protein synthesis, for example, takes place by the same reaction in all cells but the extent is less in brain or muscle than in pancreas or liver. Glycolysis and the tricarboxylic acid cycle follow the same pathway in all tissues, but the rate varies greatly. In skeletal muscle, for example, glycolysis can be very high, whereas in heart or brain it is always low.

(ii) *Control of Metabolism—Concentration of enzymes.* The rate of metabolic pathways may be determined by the amount of enzyme present; an example is the synthesis of glycogen, the reserve polysaccharide of higher animals. The key enzyme is UDP glucosyl transferase which adds glucose units on to growing polysaccharide chains. The enzyme is high in those tissues which synthesize much glycogen and low in those which do not (Table 4-1). The concentrations of some of the enzymes concerned with

Table 4-1 The distribution of glycogen synthetase in rat tissues. Tissues that store glycogen, such as muscle, liver and heart, contain the highest level of enzyme[a]

Tissue	Enzyme activity[b]
Muscle	220
Liver	187
Heart	166
Brain	32
Spleen	37
Kidney	31
Lung	36

Glycogen synthetase:
UDP glucose + acceptor → glucose–acceptor + UDP

[a] From Leloir, Olavarria, Goldemberg and Carminatti (1959). *Arch. Biochem. Biophys.*, **81**, 508
[b] μmoles UDP formed/hr/g

glycolysis and gluconeogenesis in developing chicks and young rats have already been seen to vary. But often, as in glycolysis or the tricarboxylic acid cycle of adult organs, the amount of enzyme is not the rate limiting factor. It is rather the supply of substrate or the removal of product or the extent to which enzymes are activated or inhibited by specific compounds that controls the rate.

Activity of enzymes. Control of metabolic rate by variation in activity of enzymes is illustrated in the case of muscle. The amount of glucose entering the cell, which varies according to the amount of insulin in the blood stream (see Section 5-1(c)), influences the overall rate; so does the intracellular concentration of ATP which is both a product of one reaction, glyceraldehyde-3-phosphate dehydrogenase and a substrate and inhibitor of another, phosphofructokinase (PFK) (No 4-1). AMP is an activator of this enzyme. These controls do not operate simultaneously; only one stage of the sequence is normally rate limiting and hence susceptible to control. But as conditions change, so different reactions become potentially rate limiting.

The synthesis and breakdown of amino acids, fats and nucleotides are all controlled by activation or inhibition of particular key enzymes. It is not yet known how protein synthesis is controlled in every situation (see Section 8-1); but certainly the supply of the different forms of RNA (tRNA, rRNA and mRNA), which may be considered either as substrate or activator, affects the rate.

The control of metabolism by enzyme activation, in which output responds instantaneously to input, has been termed 'fine control', in contrast to the 'coarse control' achieved by alteration in the rate of enzyme synthesis, in which there is inevitably some lag. The very responsiveness of fine control makes it difficult to see how a metabolic pathway is maintained at a specific rate in different tissues. An answer, which is only just emerging, appears to be that certain enzymes have different properties with respect to inhibition or activation by metabolites and hormones (Section 5-1(c)) in different tissues. They are called 'isoenzymes'.

(iii) *Adaptation in Differentiated Tissues.* Another problem is posed by the fact that adaptive changes (Section 1-1(b)) occur in differentiated tissues; indeed, loss of ability to adapt is a symptom of dedifferentiation (Section 6-2(a)). Depending on the state of nutrition of an animal, for example, gluconeogenesis in liver may be turned on or off; the glycolytic rate of muscle increases during exercise and falls again at rest. The mechanism by which adaptation is achieved is by variation of the activity, or synthesis, of enzymes. In other words, by the same controls that have just been described with respect to differentiation. How, then, are the basic patterns of fine and coarse control stabilized in differentiated tissues while yet allowing reversible variation to occur adaptively?

In the case of fine control, the dilemma may be resolved by postulating different *limits* within which the activity of key enzymes can fluctuate. An example which supports this view is that of lactate dehydrogenase. In skeletal muscle, the activity of the enzyme varies widely according to the environmental circumstances; in heart muscle, on the other hand, the activity is restricted to quite low levels. The basis for this difference is examined below.

In the case of coarse control. limitation is probably at the genetic level, as discussed in Section 8-1. That is to say, certain genes or their products become permanently activated or repressed in differentiated tissues. The haemoglobin genes, for example, are silent in all but erythropoietic (concerned with forming erythrocytes or red blood cells) cells, whereas the genes for glycolysis are active in virtually every type of cell. If the extent to which genes or their products are expressed varies from tissue to tissue, stable patterns of differing enzyme concentration are obtained. By allowing gene expression to fluctuate between particular limits according to external stimuli, adaptive enzyme synthesis in certain tissues may be brought about. The nature of the molecules which regulate gene expression remains to be elucidated (see Section 8-1(b)).

(iv) *Isoenzymes—Lactate dehydrogenase.* Some years ago it was noticed that when a preparation of lactate dehydrogenase (Figure 4-1) is submitted to electrophoresis on starch gel, five peaks of activity are obtained. What is particularly interesting is that the

distribution of activity in the heart, kidney and brain is characteristically different from that in skeletal muscle, small intestine and
liver (Figure 4-7).

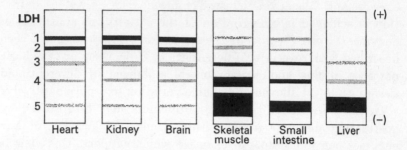

Figure 4-7 Electrophoresis of lactate dehydrogenase from various mouse
tissues. Purified extracts are subjected to starch gel electrophoresis and
enzyme activity located by a histochemical reaction. The darkness of the
bands is a measure of the amount of enzymic activity. From Markert (1965)

On separating and analysing the five peaks (called LDH-1, 2, 3,
4 and 5) it was found that each has the same molecular weight
(approximately 140,000); on the other hand, the constituent amino
acids vary, LDH-1 being richer in acidic and LDH-5 in basic
amino acids, as might be expected from their electrophoretic behaviour. The amino acids at the C and N-terminal of the protein
chain are the same. The number of such C and N-terminal amino
acids per molecule is four each, suggesting that a molecule is made
up of four sub-units of 35,000 molecular weight. This is confirmed
by other evidence. The fact that LDH preparations from many
different tissues all separate into five components, coupled with
the fact that each component has four sub-units, suggests that
there may be two types of sub-unit (called α and β) and that
LDH-1–5 are hybrids composed of all possible combinations
(Table 4-2).

That this is actually so was demonstrated by the following
experiment. The hybrids may be dissociated into sub-units (which
are enzymically inactive) by treating with 1M sodium chloride at
0°C. When the original ionic environment is restored, re-association
takes place. Pure LDH-1 or LDH-5 were treated in this way; the

Table 4-2 Sub-units of LDH isoenzymes

Sub-units	Isoenzyme
$\alpha\alpha\alpha\alpha$	LDH 1
$\alpha\alpha\alpha\beta$	LDH 2
$\alpha\alpha\beta\beta$	LDH 3
$\alpha\beta\beta\beta$	LDH 4
$\beta\beta\beta\beta$	LDH 5

resulting hybrids are identical with the starting material. When equimolar amounts of pure LDH-1 and LDH-5 were mixed, a mixture of five hybrids was obtained. They correspond in electrophoretic mobility to LDH-1–5; the amounts of the different hybrids formed was 6 per cent, 25 per cent, 38 per cent, 25 per cent and 6 per cent of the total, which is just about the ratio of 1:4:6:4:1 predicted for random reassociation of all possible forms from 4α and 4β sub-units.

The ability of sub-units to associate spontaneously in random manner implies that the relative amounts of α and β synthesized in different tissues are fixed. In other words there are two genes, one coding for the α and one for the β sub-unit of lactate dehydrogenase (as in haemoglobin, where one gene codes for the two α-chains and one for the two β-chains). Each gene is active, but to a different extent in different tissues. On reflexion, therefore, metabolic regulation by isoenzymes is not so different from that by differential synthesis of the same enzyme: each involves the controlled expression of specific genes in different tissues.

The physiological significance of LDH-1 and LDH-5 may be related to their sensitivity to pyruvate. LDH-1 is inhibited by relatively low concentrations of pyruvate, whereas LDH-5 is not. (Although the difference between LDH-1 and LDH-5 is more marked at 25° than 37°, it is still appreciable at the higher, physiological temperature of mammals.) In glycolysis, the enzyme catalyses the reaction pyruvate to lactate. If the rate of glycolysis is increased, more pyruvate is formed. It can be converted to lactate by LDH-5 but not by LDH-1 which is inhibited by its own substrate (Figure 4-8). Tissues which contain LDH-5, such as skeletal muscle, can therefore continue to produce ATP (required for muscle contraction) when the oxygen supply becomes inadequate, as in sprinting and other violent exercise. Of course, by degrading glucose anaerobically to lactate instead of aerobically to carbon dioxide and water, much less ATP is formed (2 instead of 38 molecules per molecule of glucose) and hence the rate of glucose breakdown increases in compensation. Now a high concentration of lactate, which is acidic, would be injurious to tissues such as heart or brain; it is avoided by possession of LDH-1, which becomes inhibited at the increased concentration of pyruvate accompanying rapid glycolysis.

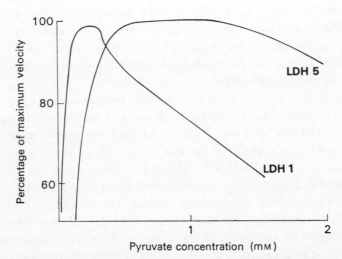

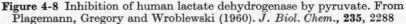

Figure 4-8 Inhibition of human lactate dehydrogenase by pyruvate. From Plagemann, Gregory and Wroblewski (1960). *J. Biol. Chem.*, **235**, 2288

Whether this interpretation of the difference between LDH-1 and LDH-5 is the only one is not known. Certainly it is true that in general organs which tolerate anaerobiosis, such as skeletal muscle, intestine and liver have high LDH-5 and those which do not, such as heart, brain and kidney, have high LDH-1.

Embryonic rat tissues are capable of a high rate of anaerobic glycolysis (Table 6-1) and as expected LDH-5 predominates. During development this is replaced by LDH-1 in those tissues (heart, brain, kidney) which are characterized by high LDH-1 in the adult animal (Figure 4-9). Sperm LDH (LDH-X) has been found to separate on starch gel electrophoresis into a pattern of hybrids different from that of LDH-1–5. It appears to contain a unique sub-unit (γ) which is synthesized by a separate gene active only in sperm tissue. The distribution of several other iso-enzymes changes specifically during development. Variation in the pattern of plant isoenzymes during development of specific organs has already been referred to (Section 2-4(b)).

The analytical separation of the hybrids of LDH has become a useful clinical tool. This is because the pattern in human serum

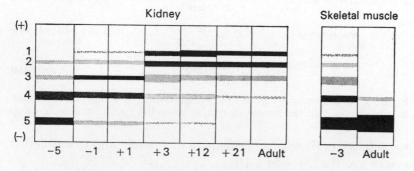
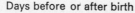
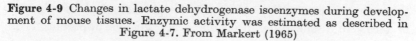

Figure 4-9 Changes in lactate dehydrogenase isoenzymes during development of mouse tissues. Enzymic activity was estimated as described in Figure 4-7. From Markert (1965)

changes in diseases, such as hepatitis or cardiac failure, which are accompanied by leakage of enzymes from the respective organs.

Other examples. Not all isoenzymes are necessarily hybrid molecules of the LDH type. So long as two or more genes are involved, enzymes with different properties can be formed. In the case of phosphorylase (Figure 4-10) for example, it is probable that the gene which codes for the muscle enzyme is distinct from that of liver, since 'glycogen storage disease V', a hereditary defect due to lack of muscle phosphorylase, does not involve the liver, while a similar disease of liver leaves muscle unaffected. Each enzyme appears to be made up of identical sub-units, as in LDH-1 or LDH-5; the point is that only one gene is active in any one cell, unlike the situation with LDH-2 to 4. Muscle phosphorylase exists in an active and an inactive form, called a and b; a is made up of four sub-units and b of two, the process of activation (see Section 5-1(c)) involving a doubling of molecular weight.

Another enzyme which varies from tissue to tissue is adenyl cyclase, the enzyme responsible for the ultimate activation of phosphorylase (Figure 4-10). In this instance it is known that the

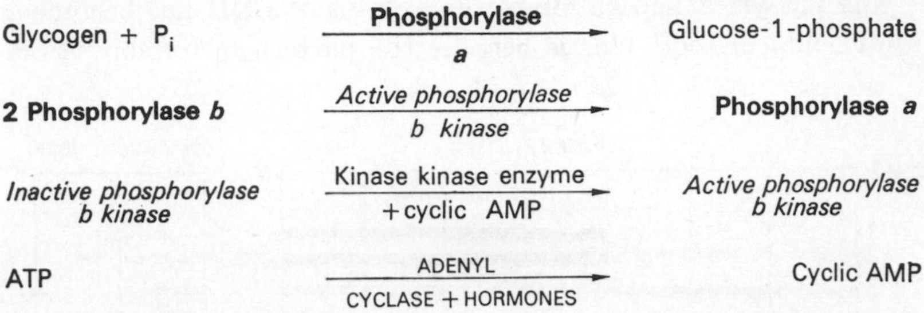

$$\text{Glycogen} + \text{P}_i \xrightarrow{\text{\textbf{Phosphorylase}} \atop a} \text{Glucose-1-phosphate}$$

$$\text{\textbf{2 Phosphorylase } } b \xrightarrow{\textit{Active phosphorylase} \atop b \text{ } kinase} \text{\textbf{Phosphorylase } } a$$

$$\textit{Inactive phosphorylase} \atop b \text{ } kinase \xrightarrow{\text{Kinase kinase enzyme} \atop +\text{cyclic AMP}} \textit{Active phosphorylase} \atop b \text{ } kinase$$

$$\text{ATP} \xrightarrow{\text{ADENYL} \atop \text{CYCLASE} + \text{HORMONES}} \text{Cyclic AMP}$$

Figure 4-10 Enzymes involved in the degradation of glycogen. Since each step is catalytic for the preceding one, a 'cascading' phenomenon is set up

enzyme responds to one hormone in one tissue, and to a different one in another (Section 5-1(c)).

Several other cases of isoenzymes have been described (for example, Section 2-4(b)). Although the physiological significance in terms of tissue differentiation is not always apparent, it is interesting that in most cases the active enzyme is made up of sub-units of similar size but slightly different chemical composition. This type of structure lends itself particularly well to a delicate kind of control in which the sub-units of a hybrid undergo conformational change in response to the action of specific activators or inhibitors. Many key enzymes in metabolism are subject to such *allosteric* regulation and it is not premature to speculate that when the rate of the pathway which they control differs, the enzymes may well turn out to be isoenzymes of the type described. The physiological importance of isoenzymes is not confined to their role in differentiation. In bacteria, for example, isoenzymes have been found to play a part in regulating the relative activity of metabolic pathways which branch off from a common point.

Evolution of differentiation. The other feature of isoenzymes is that they provide multiple phenotypes from a small number of genes. (Two genes, coding for a hybrid of 4 sub-units, give 5 different hybrids; 3 genes, coding for a hybrid of 4 sub-units, give 17). Since the sub-units of individual isoenzymes are rather similar, it is likely that the genes have a common ancestry. Presumably in the most primitive microbial cells, one gene sufficed. As cells developed diverse functions, gene duplication (see Section 8-1(*b*)) and mutation may have led to enzymes composed of multiple sub-units coded by separate genes; this allows the more subtle, allosteric control referred to above, and pertains to certain enzymes of 'current' bacteria such as *E. coli*, as well as in proteins of higher animals, such as haemoglobin in which the proportions of sub-units are fixed. As cells developed into multicellular organisms, a mechanism for controlling the relative expression of genes in different cells appears to have evolved; this enables distinct enzyme patterns to be maintained without an increase in genetic content.

4-3 Adult Morphogenesis

(a) Red Cells

Tissues develop not only during the growth of an animal but also once it has reached maturity. An example is the continuous production of specialized red blood cells (erythrocytes) from precursor or stem cells in the bone marrow. This occurs throughout the life of the animal, the human red blood cell having an average life span of some 120 days. In mammals the red blood cell contains no nucleus, no mitochondria and no endoplasmic reticulum. It cannot synthesize protein, RNA or DNA; it cannot divide, hence the need to produce more cells from nucleated precursor cells when red blood cells die. The energy for what few endergonic (energy requiring) processes the red blood cell does carry out, such as the maintenance of correct intracellular ionic concentration, is derived from anaerobic glycolysis. In birds the situation is similar; although there is a nucleus it is quite inactive and does not synthesize DNA and the cells do not divide. In many instances, including plants, it has been found that the more differentiated a cell becomes the less it undergoes mitosis (cell division) (see Section 6-1).

The sole function of the red cell is to transport oxygen from lung to tissues, which it does by combining oxygen with the specific protein haemoglobin. Haemoglobin synthesis takes place in the reticulocyte, which is a cell intermediate between a stem cell and the erythrocyte. Mammalian reticulocytes have no nucleus (and hence do not divide) but contain mitochondria and endoplasmic reticulum. Haemoglobin is about the only protein which the reticulocyte makes and the system is therefore a useful one for the study of specific protein synthesis. However the view that haemoglobin synthesis in the reticulocyte is a good example of the emergence of a differentiated system may be mistaken, since the erythroblasts, which are precursors of reticulocytes, already make haemoglobin. The developmental process is rather a loss of ability to make all other proteins, coupled with a gradual loss of intracellular structures and their associated functions. Amplification of the haemoglobin genes may account for the increased synthesis of haemoglobin. The initiation of haemoglobin synthesis occurs in the erythroblast and is stimulated by a hormone known as erythropoietin (Section 5-1(b)).

The type of haemoglobin which is synthesized by embryonic bone marrow differs from that made by the erythroblasts of adults. Adult haemoglobin is made up of four sub-units: two α-chains (141 amino acids) and two β-chains (146 amino acids). Foetal haemoglobin has two α-chains and two γ-chains (146 amino acids, similar in many regions to β-chains). The physiological significance of the two types of haemoglobin lies in the greater affinity of foetal haemoglobin for oxygen, thus ensuring efficient transfer of oxygen from the maternal circulation to the embryonic circulation in the placenta. Haemoglobin synthesis is thus controlled by three genes, only two of which are ever active at one time. The similarity to the α, β and γ genes of LDH is striking. Unlike the case of lactate dehydrogenase, the two active genes α and β, or α and γ, are expressed to an exactly equal extent. However, rare diseases occur in which this is not true and in which adult haemoglobin contains more α-chains than β. There is also a disease (thalassaemia) in which foetal haemoglobin continues to be made after birth instead of adult haemoglobin.

Other developmental events in adults are the production of white blood cells (see below) and sperms, the maturation of oocytes and many other changes associated with the oestrous cycle and with pregnancy. Hormones have been shown to trigger many of these events.

(b) White Cells

The formation of white cells or leucocytes (which comprise lymphocytes, macrophages and other cells) is much less clear than that of red cells. It is not known whether all mature cells develop from the same precursor (or stem) cell or from different ones; development from the precursor of the erythroblast has even been suggested. What is known is that the thymus is intimately involved in the synthesis of lymphocytes. The chief part which white cells play is in defending the animal against infectious agents such as bacteria and viruses. This occurs by phagocytosis or engulfment of the invading agent. Macrophages are the main cells responsible, and their number increases greatly during infection. At the same time, the blood is found to contain special proteins called antibodies, which neutralize specific structures in the bacteria or viruses, called antigens. Antigens are generally proteins; certain lipids and carbohydrates are also antigenically active. Neutralization by antibodies, followed by phagocytosis leads to the death of the microbes (Figure 4-11).

(i) *Nature of Antigens and Antibodies.* Antibodies are synthesized by certain white cells in organs such as the spleen, bone-marrow, lung and lymph-nodes. It appears that the presence of an antigen stimulates small lymphocytes to divide and confers on them the potential to produce an antibody. Although this potential is retained in the small lymphocyte (now having the properties of a 'memory cell'), the expression of antibody synthesis probably takes place in a related ('plasma') cell, which is derived from, or receives information from, the small lymphocyte. Eventually antibodies are found free in the blood stream as γ-globulins (Figure 4-11).

The most striking fact about γ-globulin production is that any one cell makes only one type of antibody. Now antibodies have been shown to vary not only with respect to the antigen that elicits their synthesis, but also with respect to certain inherited allelic characteristics, each allele in a heterozygous individual being expressed in a different cell; this phenomenon is known as

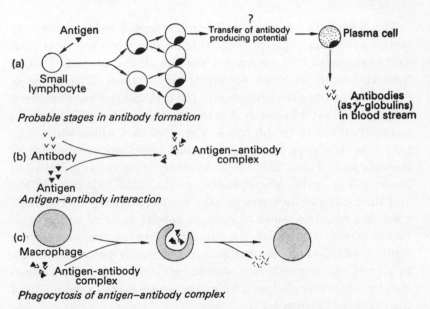

Figure 4-11 Host defence mechanism: neutralization of antigen by antibody

allelic exclusion. Antibody production thus provides one of the most clear cut examples of the synthesis of cell-specific proteins. The mechanism by which the structure of an antibody is specified remains to be elucidated. Current speculation is that genes for all possible kinds of antibody are present in lymphocytes from birth onwards and that they arise by somatic mutation (Section 6-1) of a very few basic genes present in the germ cells.

The formation of an antibody–antigen complex, which can be

caused to precipitate under appropriate conditions, provides a basis for the detection of specific antigens. A known antigen is injected into a recipient animal, generally a rabbit, and some time later the serum of the animal is collected. If this gives rise to a precipitate when mixed with a suspected antigen, its antigenic nature is confirmed. The relative similarity of different antigens can be estimated by this means (Figure 4-12). The precipitation reaction is generally

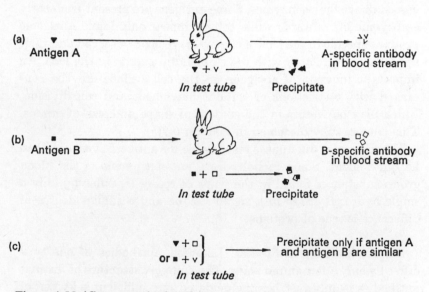

Figure 4-12 'Cross-reaction' to determine similarity between two antigens

carried out in agar as this enables small amounts of antigen to be detected. The value of the antigen–antibody reaction is that it provides an extremely sensitive and specific assay for the presence of particular proteins. It has proved especially useful for the detection of non-enzymatic proteins such as haemoglobin or myosin (for example, Section 4-1(a)).

(ii) *Cell-specific Antigens.* With this immunological technique it has been found that proteins differ not only from one individual to another, but also between tissues of the same individual. This is why antigens and antibodies are of such relevance to a discussion of differentiation. It has already been stated that the myosin of skeletal muscle is antigenically distinguishable from that of heart. Red blood cells and various secretory tissues contain a variety of carbohydrate antigens. Antigens specific for chick spleen, brain and heart have been demonstrated; indeed, each differentiated tissue probably contains specific antigens. The chemical variation between antigens such as heart and skeletal muscle myosin may be slight enough to indicate a common 'ancestry' similar to that discussed above for isoenzymes. Some antigens are present from early embryonic life onwards while others appear only later. Antigenic proteins specific to the lens of the eye have been observed and their appearance correlated with developmental events in the lens. An important function of antigens on the cell surface may be concerned with restrictions of cellular movement and cell division, which may be factors in the control of shape and size of organs. This is more fully discussed in Section 6-2(*b*).

The molecular differences responsible for antigenic variation can be quite slight. Some carbohydrate antigens, such as the blood group substances found in the walls of red cells, differ by only a single hydroxyl group in a sugar residue and a similar degree of difference is true of proteins.

(iii) *Evolution of Antibodies.* Likewise, antibodies of one type differ by only a few amino acids. The complex structure of distinct antibodies is only just being elucidated but sufficient is known of the *similarities* between different antibodies to narrow down the site at which *variation* between the many thousands of distinct antibodies known to exist in a single individual occurs.

The fact that certain regions within an antibody molecule have repetitive amino acid sequences suggests that antibodies may have evolved from smaller structures by gene duplication followed by mutation. A similar mechanism was proposed for the development of isoenzymes and other proteins (see above), though in that case the duplicated genes code for separate polypeptide chains. Since gene amplification may play a part in the development of differentiated structures within an organism (such as haemoglobin synthesis in reticulocytes and ribosomal RNA synthesis in oocytes), the old idea (Section 3-1) that embryological development and evolution have a similar basis is revived.

4-4 Concluding Remarks

This chapter has provided several clear-cut examples of the way in which variation of structural proteins and enzymes underlies the process of differentiation. In some instances, variation was seen to be achieved by an alteration in the rate of synthesis (between zero and certain upper limits), whereas in other cases it is due to differing susceptibility of isoenzymes to low molecular weight compounds. Another property which distinguishes enzymes of different tissues is their sensitivity to hormones, which is discussed in the next chapter. Since susceptibility of isoenzymes to molecular control is a function of the relative amounts of the constituent sub-units, the rate of protein synthesis becomes of paramount importance. In other words, the extent of gene expression again emerges as the controlling factor in differentiation.

The study of embryological development in cultured cells which was briefly discussed in this chapter may prove to provide important information regarding the type of molecules which are involved in gene regulation. Moreover, such systems, in which a morphological response is elicited by a specific inducer *in vitro*, may prove useful for studying the regional emergence of differentiated structure and function as mentioned in Section 3-4.

4-5 Summary

(a) Early Development

Haemoglobin and *myosin* are two *cell-specific proteins* that appear very early on during the embryonic development of chicks. The cellular activity of many enzymes subsequently changes as a result of altered *synthesis* or *activity*. In the case of phosphofructokinase and fructose-1,6-diphosphatase, which respectively rise and fall on hatching, the changes account for the observed switch from gluconeogenesis to glycolysis.

In the new-born rat, which changes from glycolysis to gluconeogenesis as lactation starts and back to glycolysis again on weaning,
the concentration of key enzymes such as fructose-1,6-diphos-
phatase alters in the expected direction. The *development* of
embryonic tissues such as cartilage, kidney tubules or pancreas can
be studied in *culture*. *Induction* of differentiation is dependent on
the presence of various *embryonic factors* which have not yet been
identified.

(b) Completed Organs

Several distinctive properties of completed organs are explicable
in terms of specific *structural proteins*. For example, collagen gives
to cartilage its tensile strength, actin and myosin enable muscles
to contract and haemoglobin renders blood cells capable of carrying oxygen from lung to tissue. In some cases the functional
capacity of completed organs is reflected by the presence of cell-
specific *enzymes*, as in the formation of glucose, ketone bodies,
urea and other products by liver.

Differences in *enzyme concentration* account for cases in which
several tissues carry out the same process but at different rates, as
in the *formation* of *glycogen*. Alternatively, a difference in the
sensitivity of enzymes—isoenzymes—to activation or inhibition
may be involved. *Isoenzymes* are generally *hybrid molecules* composed of several sub-units; this may explain their selective susceptibility to inhibition (and activation) by specific molecules.
The inhibition by pyruvate of LDH-1 accounts for the low rate of
anaerobic glycolysis in tissues which contain it, such as *heart* and
brain. Different tissues have a characteristic distribution of isoenzymes; the pattern varies during development.

(c) Adult Morphogenesis

Some developmental processes like the *maturation* of red and
white *blood cells* occur throughout the life of an animal. Certain
white cells are concerned with the formation of *antibodies*. These
are *specific proteins* made in response to the presence of an *antigen*
which can be protein, lipid or carbohydrate. Only one kind of antibody is made by any one antibody-producing cell. *Different organs,*

such as brain, spleen, heart and lens contain *antigens* specific to their own cells. In addition, some proteins, such as myosin, are antigenically different in tissues such as heart and skeletal muscle.

(d) The Basis of Differentiation

Since the activity of isoenzymes is specified by the relative amounts of component sub-units which are present, the *mechanism* for achieving *cellular variation* is essentially the same whether *altered sensitivity* or *altered synthesis* of *enzymes* is involved: namely the *differential expression* of *particular genes*.

Selected Bibliography

General Reading

Brachet, J. and A. E. Mirsky (Ed.), *The Cell*. Vol. 4 (1960), (Intr.) Vol. 5 (1961) and Vol. 6 (1964). Academic Press, London–New York
Haynes, R. H. and P. C. Hanawalt (Intr.) (1968). *The Molecular Basis of Life. Part I, Macromolecules*. Readings from *Scientific American*. W. H. Freeman & Co., San Francisco, California

Books of general tissue biochemistry

Bartley, W., L. M. Birt and P. Banks (1968). *The Biochemistry of the Tissues*. John Wiley & Sons, London
White, A., P. Handler and E. L. Smith (1968). *Principles of Biochemistry*, 4th Edition, McGraw-Hill, New York

Proteins in Early Development

Chapeville, F. and P. Fromageot (1967). 'Vestigial' Enzymes during Embryonic Development. *Adv. Enzyme Regulation*, 5, 155
Ebert, J. D. (1965). An Analysis of the Synthesis and Distribution of the Contractile Protein, Myosin, in the Development of the Heart. In *Molecular and Cellular Aspects of Development* (Ed. E. Bell), Harper & Row, New York, p. 351
Moog, F. (1967). Enzyme Development in Relation to Functional Differentiation. In *Biochemistry of Animal Development* (Ed. R. Weber), Academic Press, London–New York, Vol. 1, p. 307
Solomon, J. B. (1967). Development of Non-enzymatic Proteins in Relation to Functional Differentiation. In *Biochemistry of Animal Development* (Ed. R. Weber). Academic Press, London–New York, Vol. 1, p. 368
Vernon, R. G. and D. G. Walker (1968). Changes in Activity of some Enzymes Involved in Glucose Utilization and Formation in Developing Rat Liver. *Biochem. J.*, 106, 321

Wallace, J. C. and E. A. Newsholme (1967). Comparison of the Properties of Fructose 1:6 Diphosphatase, and the Activities of other Key Enzymes of Carbohydrate Metabolism, in the Livers of Embryonic and Adult Rat, Sheep and Domestic Fowl. *Biochem. J.*, **104**, 378

Wilt, F. H. (1965). Regulation of the Initiation of Chick Embryo Haemoglobin Synthesis. *J. Mol. Biol.*, **12**, 331

Culture Studies; Induction

Grobstein, C. (1967). The Problem of the Chemical Nature of Embryonic Inducers. In *Cell Differentiation, CIBA Foundation Symposium* (Ed. A. V. S. de Reuck and J. Knight), J. & A. Churchill, London, p. 131

Lash, J. W., F. A. Hommes and F. Zilliken (1965). The *in vitro* Induction of Vertebral Cartilage with a Low-Molecular-Weight Tissue Component. In *Molecular and Cellular Aspects of Development* (Ed. E. Bell), Harper & Row, New York, p. 95

Parsa, I., W. H. Marsh and P. J. Fitzgerald (1969). Organ Culture of the Rat Pancreas Anlage as a Model for the Study of Differentiation; I. Techniques and Comparison of *in utero* (embryonic) and *in vitro* (organ culture) morphogenesis. In *Nucleic Acid Metabolism, Cell Differentiation and Cancer Growth* (Ed. E. V. Cowdry and S. Xeno), Pergamon Press, Oxford, p. 287

Rutter, W. J., N. K. Wessells and C. Grobstein (1965). Control of Specific Synthesis in the Developing Pancreas. In *Molecular and Cellular Aspects of Development* (Ed. E. Bell), Harper & Row, New York, p. 381

Thorp, F. K. and A. Dorfman (1968). Differentiation of Connective Tissue. *Current Topics in Dev. Biol.*, **2**, 15

Tiedemann, H. (1967). Biochemical Aspects of Primary Induction and Determination. In *The Biochemistry of Animal Development* (Ed. R. Weber), Academic Press, London–New York, Vol. 2, p. 4

Wessels, N. K. and W. J. Rutter (1969). Phases in Cell Differentiation. *Scientific American*, **220**, No. 3, p. 36

Yamada, T. (1967a). Cellular Synthetic Activities in Induction of Tissue Transformation. In *Cell Differentiation. CIBA Foundation Symposium* (Ed. A. V. S. de Reuck and J. Knight), J. & A. Churchill, London, p. 116

Yamada, T. (1967b). Factors of Embryonic Induction. In *Comprehensive Biochemistry* (Ed. M. Florkin and E. H. Stotz), Elsevier Publishing Co., Amsterdam, Vol. 28, p. 113

Proteins in Completed Organs

Gross, J. (1961). Collagen. *Scientific American*, **204**, No. 5, p. 120

Huxley, H. E. (1965). The Mechanism of Muscle Contraction. *Scientific American*, **213**, No. 6, p. 18

Kühn, K. (1969). The Structure of Collagen. In *Essays in Biochemistry* (Ed. P. N. Campbell and G. D. Greville), Academic Press, London–New York, Vol. 5, p. 59

Marks, P. A. and J. S. Kovach (1966). Development of Mammalian Erythroid Cells. *Current Topics in Dev. Biol.*, **1**, 213

Perutz, M. F. (1964). The Haemoglobin Molecule. *Scientific American*, **211**, No. 5, p. 64

Pon, N. G. (1964). Expressions of the Pentose Phosphate Cycle. In *Comparative Biochemistry* (Ed. M. Florkin and H. S. Mason), Academic Press, London–New York, Vol. 7, p. 1

Scrutton, M. C. and M. F. Utter (1968). The Regulation of Glycolysis and Gluconeogensis in Animal Tissues. *Ann. Rev. Biochem.*, **37**, 249

Stracher, A. and P. Dreizen (1966). Structure and Function of the Contractile Protein Myosin. In *Current Topics in Bioenergetics* (Ed. D. R. Sanadi), Academic Press, London–New York, Vol. I

Isoenzymes

Davis, C. H., L. H. Schliselfeld, D. P. Wolf, C. A. Leavitt and E. G. Krebs (1967). Interrelationships among Glycogen Phosphorylase Isoenzymes. *J. Biol. Chem.*, **242**, 4824

Latner, A. L. and A. W. Skillen (1968). *Isoenzymes in Biology and Medicine*. Academic Press, London–New York

Markert, C. L. (1965). Epigenetic Control of Specific Protein Synthesis in Differentiating Cells. In *Molecular and Cellular Aspects of Development* (Ed. E. Bell), Harper & Row, New York, p. 267

Rutter, W. J. and C. S. Wever (1965). Specific Proteins in Cytodifferentiation. In *Developmental and Metabolic Control Mechanisms and Neoplasia*. The Williams and Wilkins Company, Baltimore, Md., p. 195

Shaw, C. R. (1969). Isoenzymes: Classification, Frequency and Significance. *Intern. Rev. Cytol.*, **25**, 297

Umbarger, H. E. (1961). End-Product Inhibition of the Initial Enzyme in a Biosynthetic Sequence as a Mechanism of Feed Back Control. In *Control Mechanisms in Cellular Processes* (Ed. D. M. Bonner), The Ronald Press Company, New York, p. 67

Vesell, E. S. (Ed.) (1968). Multiple Molecular Forms of Enzymes. *Ann. N.Y. Acad. Sci.*, **151**, Art. 1

Wilkinson, J. H. (1965). *Isoenzymes*. E. & F. N. Spon Ltd, London

Antigens and Antibodies

Bittar, E. E. and N. Bittar (Ed.) (1969). Immunology and Transplantation. In *The Biological Basis of Medicine*, Academic Press, London–New York, Vol. 4, Part 2

Burnet, M. (1969). *Cellular Immunology*. Melbourne University Press, Parkville, Victoria

Coombes, R. R. A. and P. J. Lachmann (1968). Immunological Reaction at the Cell Surface. *Brit. Med. Bull.*, **24**, 113

Edelman, G. M. and W. E. Gall (1969). The Antibody Problem. *Ann. Rev. Biochem.*, **38**, 415

Haurowitz, F. (1968). *Immunochemistry and the Biosynthesis of Antibodies*. John Wiley & Sons, London–New York

Kabat, E. A. (1968). *Structural Concepts in Immunology and Immuno-chemistry*. Holt, Rinehart & Winston, New York

Nossal, G. J. V. (1964). How Cells Make Antibodies. *Scientific American*, **211**, No. 6, p. 106

Porter, R. R. (1967a). The Structure of Immunoglobulins. In *Essays in Biochemistry* (Ed. P. N. Campbell and G. D. Greville), Academic Press, London–New York, Vol. 3, p. 1

Porter, R. R. (1967b). The Structure of Antibodies. *Scientific American*, **217**, No. 4, p. 81

Pressman, D. (1967). Structural Basis of Antibody Specificity. In *The Specificity of Cell Surfaces* (Ed. B. D. Davis and L. Warren), Prentice-Hall, Englewood Cliffs, New Jersey, p. 237

Schlesinger, M. (1967). Expression of Antigens in Normal Mammalian Cells. In *Immunity, Cancer and Chemotherapy* (Ed. E. Mihich), Academic Press, London–New York, p. 281

Speirs, R. S. (1964). How Cells Attack Antigens. *Scientific American*, **210**, No. 2, p. 58

Warren, K. B. (Ed.) (1968). *Differentiation and Immunology*, Academic Press, London–New York

Watkins, W. M. (1967). Blood Group Substances. In *The Specificity of Cell Surfaces* (Ed. B. D. Davis and L. Warren), Prentice-Hall, Englewood Cliffs, New Jersey, p. 257

Yamada, T. (1967). Cellular and Subcellular Events in Wolffian Lens Regeneration. *Current Topics in Dev. Biol.*, **2**, 247

CHAPTER 5

Hormones and vitamins

The nature of hormone action, secretion by one organ or tissue to act at a different receptor site, is so fundamental a part of differentiation that its biochemistry clearly merits attention. This chapter will deal with animal hormones. Although several plant hormones, such as auxins, gibberellins and kinins exist (Figure 2-14), their specificity for a particular tissue is slight. Rather these hormones appear to stimulate growth and development throughout the whole plant, which may again reflect the less differentiated state of plants compared with animals (Section 2-5). In certain cases, such as the effect of gibberellins on germinating seeds, the action has been shown to be by stimulation of the synthesis of specific proteins (Section 2-4(*b*)). Microbial hormones have not been found. The essence of hormones appears to be to act only on differentiated organs of higher organisms. The reason for discussing vitamins is less obvious. But as will be seen, certain vitamins, the fat soluble ones in particular, share with hormones the property of tissue specificity, so that a discussion of their biochemistry is relevant to this treatise.

5-1 Hormones

(a) Nature of Hormones

Hormones have been defined as chemical messengers of the blood stream, the word being derived from the Greek ὁρμαων, 'urging on'. This concept is based on experiments in which specific organs are surgically removed. For example if the pancreatic tissue of a dog is cut out, the animal begins to show the same symptoms as human diabetics, namely muscular weakness coupled with an abnormally high concentration of sugar in blood and urine. If the pancreatic tissue is now grafted back, but at a

4

different site such as under the skin, the symptoms disappear. This experiment, originally carried out in 1892, shows that a substance —subsequently called insulin—is secreted into the blood stream by the pancreas and that it functions to control blood sugar and to keep muscle tissue healthy. The disease diabetes mellitus is characterized by a failure of the pancreas to secrete insulin.

Insulin and other hormones have been purified and their structures analysed. No common pattern has emerged. Some hormones, like insulin, growth hormone, thyrotrophic hormone, ACTH and erythropoeitin, are proteins; some, like vasopressin and oxytocin, are small polypeptides containing six to eight amino acids and some, like thyroxine, adrenaline, serotonin, acetylcholine and histamine, are derivatives of single amino acids. Others, like cortisone, aldosterone, testosterone, oestrone, progesterone and ecdysone, are steroids (Table 5-1). The specificity of hormone

Table 5-1 The chemical nature of some mammalian hormones

Proteins	Polypeptides	Amino acid derivatives	Steroids
Glucagon (m.w. 3500)	Oxytocin (octapeptide)	Thyroxine and triiodothyronine	Adrenal cortical steroids, e.g. aldosterone, cortisone and hydrocortisone
Adrenocorticotrophic hormone (ACTH) (m.w. 4500)	Vasopressin (octapeptide)	Adrenaline (=epinephrine)	Androgens, e.g. androsterone and testosterone
Insulin (m.w. 6000)		Histamine	Oestrogens, e.g. oestradiol and oestrone
Thyrotrophic hormone (m.w. 26,000–30,000)		Serotonin	(Ecdysone, in insects)
Growth hormone (m.w. 25,000–48,000)		Acetylcholine	

action has already been referred to with regard to muscle function and insulin. Several more striking examples can be cited (for example, Figure 5-1).

It is clear that hormones are necessary for the proper functioning of differentiated organs. But this is not necessarily their main

(a) Before injection   (b) After injection

Figure 5-1 Effect of male sex hormone on restoring growth of cock's comb. A capon castrated as a chick was subsequently injected with androsterone. From Callow and Parkes (1935). *Biochem. J.*, **29**, 1414

physiological action; the observed malformations are sometimes only consequences of some quite different cause. The way this comes about may be illustrated in the case of thyroxine. This hormone (Figure 5-2) is manufactured in the thyroid gland in the

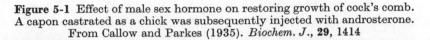

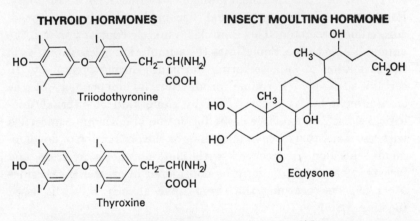

Triiodothyronine

Thyroxine

Ecdysone

Figure 5-2 Structures of some 'developmental' hormones

base of the neck. Synthesis of thyroxin is stimulated by another hormone, thyrotrophic hormone, secreted by the anterior pituitary which is a small gland situated in the brain. If the concentration of thyroxine in the blood rises above a certain limit, it cuts off further secretion of thyrotrophic hormone. This in turn switches off thyroxine production in the thyroid until the level of thyroxine returns to normal. Such a homeostatic or 'feedback' mechanism is typical of the behaviour of several hormones. Now if for some reason, such as a deficiency of iodine in the diet, the thyroid cannot produce enough thyroxine, the secretion of thyrotrophic hormone will never be shut off and will permanently stimulate the thyroid to perform a function it is unable to carry out. The result is an overgrowth of the thyroid gland, a condition known as goitre. Administration of thyroxine cures the malady. But other changes occur as well. The goitrous individual becomes sluggish in movement and in mental capacity, while at the physiological level a reduction in basal metabolic rate is one of several characteristic consequences. It is this process which is one of the primary targets of thyroxine action and which, in the absence of a thyroid gland, may be kept functioning normally by administration of thyroxine.

Hormones not only maintain, but also initiate metabolic events. A major function of many hormones is to transmit alterations in the environment so that the organism can adjust itself accordingly. For example, when an animal becomes suddenly frightened, adrenaline is secreted. This stimulates muscle contraction and the animal is able to flee. Sometimes the stimulation is less rapid, as in the formation of milk secreting hormones during pregnancy. In certain situations, hormone production is independent of the environment but is triggered off by some kind of internal 'biological clock'. An example is the formation of hormones concerned with the menstrual cycle in mammals or the production of developmental hormones concerned with metamorphosis in frogs and insects. Metamorphosis may be considered in greater detail since other aspects regarding the event have already been discussed (Section 3-3(b)).

(b) Hormonally Induced Synthesis of Enzymes

(i) *Metamorphosis in Frogs and Insects*. The hormones involved in the metamorphosis of tadpoles into frogs are thyroxine and the metabolically related triiodothyronine (Figure 5.2). If thyroxine is fed to young tadpoles, they stop growing and instead undergo precocious development into frogs. If the thyroid gland of young tadpoles is removed, they fail to metamorphose and develop into giant tadpoles; on receiving thyroxine the giants commence metamorphosis. The production of thyroxin in the thyroid has been followed by feeding young tadpoles with radioactive iodine. This becomes incorporated by the thyroid into thyroglobulin which in turn gives rise to thyroxine. The appearance of thyroxine begins just as the precritical stage (not yet committed to metamorphosis) gives way to the critical. Soon after this, specific enzyme synthesis, for example the synthesis of urea cycle enzymes and the protease involved in tail resorption (Section 3-3(*b*)), begins. Another developmental event which is dependent on thyroxine is the switch from synthesizing embryonic haemoglobin (Section 4-3(*a*)) to synthesizing adult haemoglobin. The production of thyroxine continues throughout the life of the adult frog but its actions are not clear.

One of the major developmental changes in insects is moulting, which occurs during the formation of both pupa (pupation) and adult (Figure 5-3). One of the agents responsible, termed ecdysone,

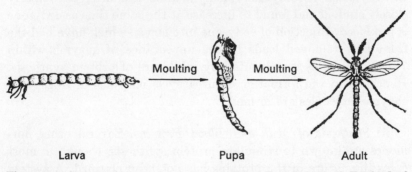

Larva   Pupa   Adult

Figure 5-3 Stages of insect development. From Ross (1965). *A Textbook of Entomology*, 3rd edition, John Wiley and Sons, New York, p. 379

has been detected in the front part of the larva prior to pupation; if this is removed, the remainder fails to develop. Injection of ecdysone causes pupation within twenty-four hours. The structure of ecdysone has recently been found to be that of a steroid (Figure

5-2). One of the characteristic changes during moulting is the darkening and hardening of the larval cuticle; this is due to the incorporation of compounds into the epidermal cells in a manner analogous to that seen in the hardening of plant root walls (Section 2-4(c)). In this case the compounds are oxidation products derived from tyrosine (Figure 5-4). The enzyme which catalyses the de-

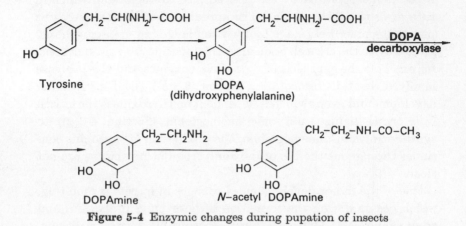

Figure 5-4 Enzymic changes during pupation of insects

carboxylation of dihydroxyphenylalanine has been particularly closely studied and found to increase at the same time as ecdysone is produced. Injection of ecdysone into larvae which have had the front part removed leads to the appearance of enzyme within 7–10 hours. This is prevented by inhibitors of protein synthesis. Hence, as in the formation of urea cycle enzymes in frogs, new enzyme molecules are formed.

(ii) *Synthesis of Milk and Blood Proteins.* Several other hormones are known to stimulate protein synthesis, though in most cases the nature of the proteins has not been clarified. A system which lends itself well to studying the relation between differentiation in terms of enzyme synthesis and the action of hormones is the lactating mammary gland. Mammary tissue taken from a pregnant mouse can be cultured rather like chick somite or rat pancreas cells (Section 4-1(c)) and the development of specialized secretory cells followed; at the same time specific milk proteins

such as casein and the enzyme system catalysing the synthesis of
the milk sugar lactose appear. This involves the formation of new
molecules, since inhibitors of protein synthesis prevent their appearance. The presence of three hormones, insulin, hydrocortisone
and prolactin, is necessary. Insulin is required to initiate the DNA
synthesis that appears to be a prerequisite for the induction of
protein synthesis. The part played by the other hormones is not
clear and may partly reflect some limitation of the conditions
in vitro compared with those in the whole animal.

Another system in which induction of a specific protein by
hormones has been studied is foetal rat liver in culture. Given
erythropoietin, haemoglobin synthesis takes place. Inhibitors of
protein synthesis abolish the induction (in adults, the capacity to
make haemoglobin is transferred to the erythroblast cells of bone
marrow, as discussed in Section 4-3(*a*); note that this change, at
least in frogs, is another event controlled by thyroxine).

Hormonal induction of enzymes is frequently accompanied by
increased RNA synthesis (Table 5-3). This has given rise to the
view that the mechanism is by a stimulation of the synthesis of
the relevant mRNA molecules (see Section 5-1(*e*)).

The effect of throxine and ecdysone in initiating the synthesis
of enzymes characteristic of particular developmental events is so
striking that it has given rise to the view that hormones are the
primary agents which cause differentiation in higher animals.
However, as will be seen below (Section 5-1(*d*)) this is an untenable
hypothesis.

So far, hormones that stimulate the *synthesis* of specific enzymes
have been considered. Several hormones are known to *activate*
existing enzymes—the fine as opposed to the coarse control of
Section 4-2(*b*)—and two examples of this may now be examined.

(c) *Hormonally Induced Activation of Enzymes*

(i) *Sugar Transport*. First of all it must be made clear that the
transport of substances in and out of cells is often an enzymic
process. This is well illustrated in the case of glucose.

Muscle cells are relatively impermeable to glucose (Figure 5-5).
The concentration in blood may be very high but that in muscle

remains low; some sugars hardly enter muscle at all. That the uptake of glucose is an enzymic process (termed 'facilitated' diffusion) is indicated by the following facts:

(1) The uptake by isolated slices of muscle tissue is temperature-dependent, being approximately twice as fast at 37° as at 27°.

(2) It is linked to energy expenditure; inhibition of metabolic reactions leading to the production of ATP, such as poisoning by arsenicals or cyanide, prevents uptake. This may be an indirect effect, reflecting variations of intracellular glucose concentration.

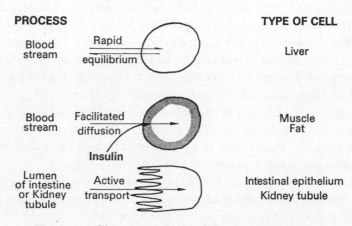

Figure 5-5 Glucose uptake by different types of cell

(3) It is dependent on the external concentration of glucose, the rate of uptake increasing to a maximum as the external concentration rises. This implies the existence of some binding mechanism which becomes saturated at high concentrations, a phenomenon characteristic of all enzyme-catalysed reactions.

(4) The uptake mechanism as expressed by its binding properties is very specific; the uptake of D-glucose, D-galactose and D-xylose, for example, is 50 per cent saturated at quite different concentrations of sugar. These concentrations are sometimes referred to as Michaelis constants (K_m) by analogy with enzymes. It is the affinity for the 'carrier' substance that appears to be stimulated by insulin; that is to say the K_m is decreased. So far all

attempts to isolate the carrier have failed; definitive proof of its enzymic nature awaits a technological advance in the manipulation of 'structural' enzymes in general. The uptake mechanism of glucose by adipose (fat) cells is, like muscle, by insulin-sensitive facilitated diffusion. Uptake by red blood cells is also by facilitated diffusion but is not stimulated by insulin.

In contrast glucose enters liver cells very easily and rapidly, most probably again by facilitated diffusion. In this instance, the intracellular concentration is a reflection of the level in the blood stream. This is because the liver is particularly well perfused with blood vessels, as is apparent from its red colour, and because its cells have rather permeable membranes. No particular barrier is involved and insulin has no effect.

A third type of situation exists in intestinal and kidney cells. As in muscle, the uptake process is enzymic. Unlike muscle, the concentration within a cell may exceed that outside several-fold; it must therefore be linked to energy-producing systems on thermodynamic grounds alone. The process also differs from that of muscle or adipose tissue by being insensitive to the action of insulin. In other words, dietary glucose is absorbed from the gut whenever it is present and it is prevented from being lost in the urine by continual reabsorption in the kidneys. The three systems provide a good illustration of the differentiation of plasma membranes, so far as glucose transport is concerned.

Returning to the effect of insulin on muscle cells, the fact that the K_m of the carrier is altered suggests an effect on its activity, rather than on its synthesis. Moreover, the action of insulin is almost instantaneous and is observed in the absence of protein synthesis. Insulin stimulates not only the transport of glucose and related sugars into muscle and fat cells but also that of amino acids; the process has been shown to be enzymic by the criteria listed above. The carrier, which is distinct from that involved in sugar transport, is again activated by insulin in the presence of inhibitors of protein synthesis.

The effect of insulin on isolated muscle tissue accounts quite satisfactorily for the observed changes, namely lowering of blood sugar, in the whole animal. However, on closer examination it is found that insulin promotes additional changes in different organs. In liver, for example, insulin increases protein synthesis as measured by incorporation of radioactively labelled amino acids. This is probably a reflexion of the synthesis of some specific proteins such as a kinase enzyme for the phosphorylation of glucose or the enzyme concerned in the transfer of glucose units to growing glycogen chains, though in the latter case activation of existing enzyme (from dependence on glucose-6-phosphate to independence) may be a contributing factor.

(ii) *Cyclic AMP*. Reference was made earlier in this chapter to the action of adrenaline in stimulating muscle contraction in animals when frightened. Accompanying this is a breakdown of glycogen to glucose, which is further metabolized to provide extra energy. The enzyme responsible for glycogen breakdown is phosphorylase (Figure 4-10).

Muscle phosphorylase, which has been particularly extensively studied in the rabbit, exists in two forms (see Section 4-2(*b*)); *a* is the one which is physiologically active (Figure 4-10). The conversion of *b* to *a* is catalysed by an enzyme (phosphorylase *b* kinase) which itself exists in an inactive and an active form, the interconversion by a kinase kinase taking place in the presence of ATP and cyclic AMP (Figure 2-10). Probably ATP is split to ADP, and inorganic phosphate attached to the enzyme as in phosphorylase *b* → *a*, but the stoichiometry is not yet known. The production of cyclic AMP is itself catalysed by an enzyme, adenyl cyclase. It is this enzyme which is sensitive to the action of adrenaline. Since each stage is catalytic for the preceding one, only a small amount of cyclic AMP is required to stimulate the degradation of bulk quantities of glycogen. This phenomenon has been referred to as 'cascading'.

The situation in liver is slightly different in that the physiologically active phosphorylase is apparently formed from an inactive phosphorylase without a change in molecular weight, but is similar in that the activation is again catalysed by a kinase which is itself activated by a kinase which is itself activated by cyclic AMP which is formed from ATP when stimulated by adrenaline.

In this case another hormone, glucagon, can act in place of adrenaline.

Many organs contain the adenyl cyclase system and respond to various hormones (Table 5-2). The outcome of increased cyclic

Table 5-2 Tissues in which cyclic AMP formation is stimulated by hormones[a]

Tissue	Hormone	Biochemical response	Physiological result
Liver	Adrenaline⎫ Glucagon ⎬	⎰Activation of ⎱phosphorylase Inhibition of glycogen synthetase	Glucose release
Fat	Adrenaline ⎫ Growth hormone ⎬Activation of lipase Cortisone ⎭		Fatty acid release
Muscle	Adrenaline	Activation of phosphorylase	Glucose release
Adrenal cortex	ACTH	?	Steroid production
Kidney	Vasopressin	?Activation of transporting system	Sodium uptake and water reabsorption
Thyroid	Thyrotrophic hormone	?Activation of protein synthesis	Iodine uptake; thyroxine release
Gastric mucosa	Histamine	?	Hydrochloric acid secretion
Pancreas	?Glucagon⎫ ?ACTH ⎬	?	Insulin release

[a] Adapted from Bartley, Birt and Banks (1968). *The Biochemistry of the Tissues*. John Wiley & Sons, London, p. 282.

AMP is not only the breakdown of glycogen by phosphorylase; in toad bladder and kidney, for example, the effect is an increased transport of water. In adipose tissue the effect is to cause lypolysis. That the effect of the hormones listed in Table 5-2 is through an increased production of cyclic AMP is indicated by the following facts:

(1) Inhibition of the a cyclic AMP diesterase, which destroys cyclic AMP by hydrolysis to AMP, mimics the action of the hormone; the drug theophylline acts in this way.

(2) Cyclic AMP itself elicits a response similar to the hormones.

(3) The concentration of cyclic AMP rises immediately after hormone administration. Two examples may be cited: (*a*) Cyclic AMP is increased five-fold some two to four seconds after administering adrenaline to an isolated heart preparation; this is prior even to the increase in heart beat. Phosphorylase does not become activated until some forty seconds later; hence glycogen breakdown cannot be the primary stimulus for increased heart beat, though it may well help to maintain the higher rate. How cyclic AMP affects muscle contraction in heart is not yet known. (*b*) In female rats which have had their ovaries removed, injection of the hormone oestradiol causes a doubling of cyclic AMP in the uterus within fifteen seconds. The rapidity of these responses suggests that activation of adenyl cyclase, or inhibition of the diesterase, rather than synthesis of new enzyme molecules is responsible for the increase in cyclic AMP.

The variety of biological responses elicited by cyclic AMP has caused it to be termed a 'second chemical messenger' in transmitting the primary effects of many hormones in higher organisms. In more primitive forms of life such as slime moulds (Section 2-2(*c*)) and possibly bacteria, it may itself be the primary agent responsible for alerting the organism to environmental change.

The specificity of the response of adenyl cyclase to hormones (Table 5-2) suggests that its structure is distinct in different tissues. Whether it is an isoenzyme of the type discussed in Section 4-2(*b*) or whether the enzyme system contains a cell-specific macromolecular cofactor, must await purification of the enzyme. Certainly adenyl cyclase illustrates the importance of the receptor site (see below), rather than the hormone itself, in the process of differentiation.

(d) Receptor Sites for Hormones

(i) *Importance of Receptors.* The possibility was raised in connection with the effect of hormones on inducing enzyme synthesis, that hormones may cause differentiation. However one must distinguish between *initiating* a process, which hormones undoubtedly do, and *establishing the potentiality* for the process to occur. The act of pressing a light switch causes a room to become

illuminated. But illumination can only occur if the correct wires
have previously been installed. Similarly hormones act only on
systems already programmed to respond to their presence. This is
obvious from the very existence of target organs. Thyrotrophic
hormone, for example, acts only on the thyroid gland, milk
secreting hormone stimulates only mammary gland, insulin promotes the uptake of sugar only into muscle cells and not into liver
or intestinal cells, and so on.

Equally important are the organs which synthesize and secrete
the hormone. Thyrotrophic hormone and milk secreting hormone
are synthesized only in the anterior pituitary gland, insulin in the
pancreas, thyroxine in the thyroid, and so forth. In other words, the
action of a hormone cannot be the prime cause of a developmental
process such as metamorphosis; it must have been preceded by the
synthesis of a hormone-specific receptor site, as well as by synthesis of the hormone itself.

This is well illustrated in the case of thyroxine. If the hormone is
given to a young tadpole whose thyroid has not yet developed,
metamorphosis takes place several days sooner than normal. But if
thyroxine is given to *very* young tadpoles (less than two days old in
the case of *Xenopus*), metamorphosis does not occur at all. The
receptor has clearly not yet been made. As new receptors appear,
so the action of a hormone may become modified. This is particularly true of thyroxine, which has a much less pronounced, and
different, effect on adult frogs. Another example is provided by
growth hormone, one of the protein hormones of the pituitary
gland. In the young animal such as a kitten it stimulates growth;
if fed to adult cats it causes diabetes by virtue of an interference
with the secretion of insulin by the pancreas.

A further argument against the hypothesis that the basis of
differentiation lies in the action of hormones is provided by the
cases in which differentiated cells can be cultured in the absence
of hormones without the loss of morphological or biochemical
characteristics. This is true of retinal cells of the eye (which continue to synthesize characteristic pigments) and of certain cancerous cells derived from connective tissue called mast cells which
continue to make granules rich in heparin (a sulphated mucopoly-

saccharide) and other specific substances. Since the cells have already differentiated by the time they are put in culture, these experiments rule out participation by hormones only in so far as *maintenance* of differentiation in this system is concerned.

Whether one can consider the synthesis and action of hormones as part of the *commitment* process (Chapter 1) is a moot point. On the one hand some differentiation has already taken place by the time the receptor system is complete; on the other hand the system can only express itself in the presence of the hormone. (As mentioned above, if the thyroid of young tadpoles is removed, they never metamorphose.) A more profitable topic for discussion is the *nature* of the receptor system.

(ii) *Nature of Receptors.* One idea that has been put forward is that the genes coding for the enzymes that are synthesized are the receptors. The 'puffing' of chromosomes by ecdysone has been cited to support this hypothesis. In many insects and their larvae the chromosomes of the salivary gland are very much larger than in other cells, rendering them easily visible under the microscope. This is because many hundred copies of each chromosome (known as polytene chromosomes) are present. If ecdysone is administered to insect larvae prior to pupation certain regions of the chromosome are seen to 'puff' (Figure 5-6). This puffing is said to be accompanied by gene expression, since it coincides with the synthesis of enzymes such as those concerned with moulting; moreover the synthesis of RNA—the mediator of information between genes and proteins—has been detected autoradiographically at just the sites where puffing occurs. Also ecdysone appears to be bound particularly tightly to the cell nucleus.

Whether puffing takes place in the epidermal cells which actually synthesize moulting enzymes such as dihydroxyphenylalanine decarboxylase is not known, as individual chromosomes are not readily detectable within cells; but what is clear is that the puffing in salivary gland induced by ecdysone cannot be related to the synthesis of dihydroxyphenylalanine decarboxylase in epidermis, as the enzyme is not made in salivary gland any more than it is made in heart or brain. In other words it is only the dihydroxyphenylalanine decarboxylase gene in *epidermis* (although present in every cell, as will be seen in Section 7-1) that is activated by ecdysone. This implies that whatever it is that prevents the gene from expressing itself, and proteins have been suggested as likely agents (Chapter 8), must possess tissue specificity. Such agents may prove to be important receptor sites for hormones (see below).

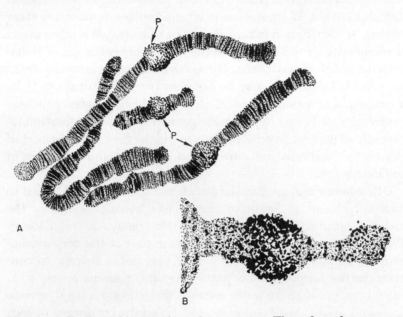

Figure 5-6 Puffing of insect giant chromosomes. These four chromosomes (A) from the salivary gland of *Chironomus tentans* larvae are each some 10 times longer and 100 times thicker than chromosomes from other tissues. Puffs (P) are seen at characteristic bands along the chromosome; the pattern varies during development. Intensive RNA synthesis accompanies puffing, as seen (B) by exposing larvae to radioactive uridine and preparing an autoradiograph (Figure 3-3). The black dots are due to radioactive RNA which is concentrated at the site of the puffs. From Beermann and Clever (1964)

As regards the physiological significance of puffing in salivary gland chromosomes, this is not yet known. It is probably a signal for some quite general events connected with metamorphosis, which bear only incidental relation to moulting enzymes in epidermis. Unless, of course, one accepts the rather unlikely view that

moulting genes are transcribed into RNA in all tissues, but trans-
lated into enzyme only in epidermal cells (the translational control
of protein synthesis, see Section 8-1). In that case one must
postulate, in addition to the hormone-sensitive gene repressor of
all tissues, a second regulator specific to epidermal cells.

One is on somewhat firmer ground if one turns from genetic
mechanisms to simple facts regarding the binding of hormones in
different tissues. If a hormone is labelled with a radioactive atom
so that its distribution in an animal can be traced, it is often found
that injection of the hormone leads to the concentration of radioactivity in the target tissue. By extracting such tissue the agent
responsible for binding may be isolated. It usually turns out to be
a protein. For example acetyl choline, which triggers muscular
contraction, is bound by a specific protein situated at the stimula-
tory site of muscle. Several hormones which effect the transport of
salts across epithelial cells are bound by a protein component of
the membrane.

Other hormones, such as the female sex hormones, are bound to
nuclear proteins of sensitive organs like uterus, vagina or the
anterior pituitary (which is the site for 'feedback' regulation).
Whether these tissue-specific proteins are part of the very mechanism of hormone action, or whether they serve merely to con-
centrate the hormone at a particular site, remains to be seen;
though one protein at least, namely the enzyme adenyl cyclase
discussed above, has been identified as a primary target of hormone action.

(e) Hormones and Membranes

It has now been seen that hormones can act in two quite distinct
ways; they can either act on the synthesis of enzymes or on their
activity. The same hormone may cause protein synthesis in one
tissue and enzyme activation in another. It is worth asking
whether it is at all possible to correlate the two effects. An answer
is indicated by investigating the time sequence of some of the
events following administration of triiodothyronine to young tadpoles. Not only protein but RNA and phospholipid synthesis is
seen to increase at an early stage, nuclear RNA being particularly

sensitive (Figure 5-7). Several other hormones (Table 5-3) show an early stimulation of nuclear RNA synthesis. This might be expected if the RNA that is formed is the 'messenger' for the pro-

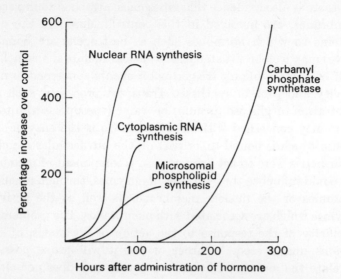

Figure 5-7 Effect of triiodothyronine on metabolic processes in premetamorphic tadpoles. From Tata (1967)

teins that are subsequently synthesized. However, on closer analysis it turns out that the RNA is actually ribosomal RNA, which subsequently presumably appears in the cytoplasm. The

Table 5-3 Hormones that stimulate ribosomal RNA synthesis in the cell nucleus of various tissues[a]

Hormone	Tissue	Animal
Growth hormone	Liver	Hypophysectomized rat
Growth hormone	Muscle	Hypophysectomized–castrated rat
Thyroid hormones	Liver	Thyroidectomized rat
Thyroid hormones	Liver	Tadpole
Testosterone	Prostate	Castrated rat
Testosterone	Liver	Castrated rat
Testosterone	Muscle	Hypophysectomized–castrated rat
Oestrogen	Uterus	Ovariectomized rat
Cortisone	Liver	Rat

[a] From Tata (1967)

exact function of ribosomal RNA in protein synthesis is not clear; but it is known that the synthesis of specific proteins can be regulated at a stage *beyond* messenger RNA formation (Chapter 8), and there is also evidence that ribosomes, attached to intracellular membranes, are involved in this control. Moreover the nuclear proteins to which hormones such as oestrogens are bound (see above) may well be situated at the nuclear membrane, which could itself be responsible for triggering off events concerned in nuclear activity and protein synthesis. Transport processes, such as the penetration of glucose (insulin) or water (vassopressin) into cells, are clearly concerned with membranes. So is the enzyme adenyl cyclase which is bound to, or part of, the intracellular membrane. If hormones were to act primarily on a component of membranes, this could influence the synthesis of enzymes, through an effect on ribosomes or the nuclear membrane, as well as the activity of enzymes which are associated with membranes. The problem of the specificity of the response to the action of hormones, of course, remains unanswered. Whether or not it will prove possible to correlate the actions of hormones in this way does not alter the fact that membranes emerge as important targets for hormone action. More than a decade ago it was suggested by R. A. Peters that the 'cytoskeleton' or internal membrane of the cell is re-orientated by the presence of hormones and that this underlies their biochemical action.

5-2 Vitamins

The term vitamin(e) was coined to describe a trace factor in food which, while not contributing to the caloric requirements of an animal, was *'essential to life'* and on purification and structural analysis turned out to be an *amine* (vitamin B_1 or thiamine). Subsequently several other factors were discovered and although they are not all amines, the term vitamin has been used for all. Lack of any vitamin gives rise to specific deficiency diseases. The vitamins, which have proved to be the same for all higher animals, have been operationally classified as water-soluble and fat-soluble. This is also a fairly useful division as far as differentiation is concerned.

(a) *The B Group*

The water-soluble ones (with the exception of vitamin C, ascorbic acid) have been termed the vitamin B complex, since most are present together in any particular source of food. The vitamins are all small organic molecules of varying structure (Figure 5-8). The function of the B group has been found to be that of coenzymes in the major pathways of metabolism; usually the active form of the vitamin is a phosphorylated or other derivative (Figure 5-8). Thus thiamine pyrophosphate is necessary for the decarboxylation of α-keto acids such as pyruvate and α-ketoglutarate; nicotinic acid and riboflavin, in the form of nucleotide derivative, are part of many dehydrogenase enzymes; pantothenic acid, as coenzyme A, is involved in the formation and metabolism of acyl groups; and so forth.

Now these sort of metabolic reactions are common to animals, plants and microbes and it may be wondered whether the same coenzymes are involved. This is exactly so, the difference being that the precursor vitamins need not be supplied intact but are synthesized by plants and microbes themselves. An exception occurs in certain *mutant* microbes which have lost the ability to synthesize a particular vitamin and require its presence in the growth medium. Isolation of mutants specific to each of the B vitamins has played a major part in elucidating the action of the vitamins.

(b) *Vitamins A, C, D, E and K*

(i) *Relevance to Differentiation.* The fat soluble vitamins (A, D, E and K) and vitamin C are also small molecules and again their structures are known (Figure 5-9). On discovering that each of the B vitamins plays a coenzymic role in metabolism it seemed natural to suppose that the same would be true of the other vitamins. However there are two clear differences. First of all these vitamins are not required by plants or microbes; attempts to produce bacterial mutants to them have failed. Secondly their action in higher organisms appears to be specific to certain tissues only, compared with the B group which act in every cell capable of catalysing the common reactions involved in the degradation,

VITAMIN

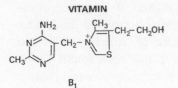

B_1

COENZYME

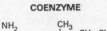
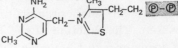

Thiamine pyrophosphate

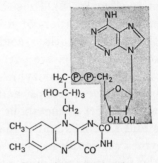

B_2

Flavin adenine dinucleotide

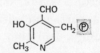

B_6

Pyridoxal phosphate

-CH

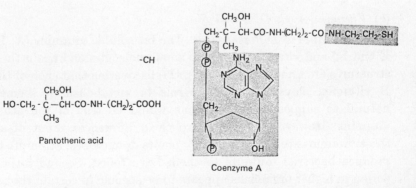

HO-CH$_2$ - C —CH-CO-NH-(CH$_2$)$_2$-COOH

CH$_3$OH

CH$_3$

Pantothenic acid

Coenzyme A

Figure 5-8 Structure of vitamins and their coenzymic forms

synthesis and interconversion of sugars, amino acids, nucleotides and fats. Thus vitamin A plays a part in vision, vitamin C in

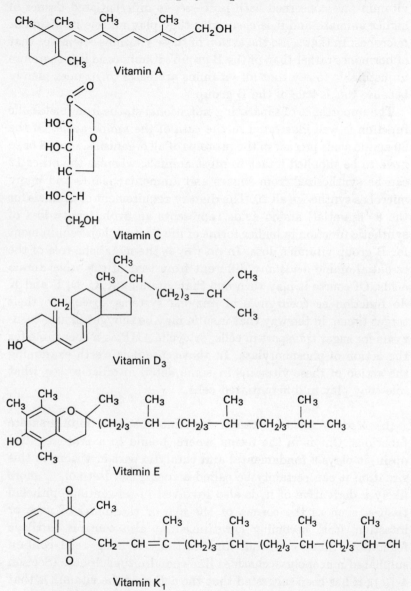

Figure 5-9 Structure of some fat-soluble vitamins and vitamin C

cartilage formation, vitamin D in bone structure, vitamin E in fertility and vitamin K in blood clotting. It is apparent that these vitamins are concerned with processes in differentiated tissues of higher animals, and it is clear why they play no role in plants or microbes. In that sense the action of these vitamins resembles that of hormones rather than of the B group vitamins and it is therefore unjustifiable to say that all vitamins are basic coenzymes merely because this is true of the B group.

The importance of separating nutritional status from metabolic function is well illustrated in the case of the amino acids. Of the 20 amino acids present in the proteins of all organisms, some 8 or so have to be supplied intact to most animals, whereas the other 12 can be synthesized from sugars and ammonia; plants and many microbes synthesize all 20. The dietary requirement of animals for the 8 'essential' amino acids represents an evolutionary loss of synthetic function in higher forms of life, just as their requirement for B group vitamins does. In no way is the metabolic role of the essential amino acids any different from that of the other amino acids. Of course it may turn out that vitamins A, C, D, E and K do function as coenzymes in enzyme systems specific to their target tissue, in the way that insulin may be thought of as a coenzyme for sugar transport in cells, or cyclic AMP as a coenzyme for the action of phosphorylase. In that sense it is worth examining the action of these vitamins in some detail in order to see what role they play in differentiated cells.

(ii) *Mode of Action*. Vitamin A, retinol, has two quite separate functions. One is in the retina where, bound to a protein called opsin, it plays a fundamental and catalytic part in vision; in this situation it can certainly be called a coenzyme. Retinol, or more likely a derivative of it, is also involved in preventing epithelial tissues, such as the cornea of the eye or the lining of nose or intestine, from becoming keratinized; it also controls cartilage formation. Since epithelial and cartilage-forming cells contain sulphated mucopolysaccharides like chondroitin sulphate (Section 4-1(c)), it has been suggested that the action of the vitamin is that of a coenzyme involved in the biosynthesis of sulphated mucopolysaccharides. Further evidence has implicated the first stage of the incorporation of inorganic sulphate into mucopolysaccharides, which is an activation reaction involving ATP, as a possible site. This has not been substantiated by recent work. Moreover the sulphate-activating enzyme is not specific to animals but is common to microbes and plants where it is concerned with the reduction of inorganic sulphate to the amino acid cysteine. Furthermore, animal cells which synthesize sulphated mucopolysaccharides, such as cancerous mast cells or cartilage-forming cells in culture, grow and make heparin (a sulphated mucopolysaccharide) in the absence of vitamin A. Hence the part played by vitamin A in epithelial and cartilage cells remains obscure. An interesting speculation that vitamin A is involved in the structure of membranes awaits experimental proof.

Vitamin C, ascorbic acid, is also concerned in cartilage formation, in this case in the biosynthesis of collagen (Section 4-2(a)). It appears to act, possibly as a coenzyme, in specific hydroxylation reactions leading to the formation of hydroxyproline and hydroxylysine which are the amino acids unique to collagen. The hydroxylation occurs after the amino acids have been activated in the process of protein synthesis.

Vitamin D, calciferol, is necessary for maintaining the structure of bones, the chief constituent of which is crystalline calcium phosphate (apatite, $Ca_{10}(PO_4)_6-(OH)_2$). Its role may be an indirect one, in that one of its biochemical actions is the stimulation of calcium absorption in the intestine.

The part played by vitamin E, tocopherol, is poorly understood. Its physiological role is concerned with fertility but all that is known biochemically is that it is an effective anti-oxidant. For that reason it has been suggested that it functions in the electron transport chain, but this is not very likely.

Vitamin K, phylloquinone, has been implicated in the synthesis of certain factors required for the so-called clotting reaction of plasma, which culminates in the precipitation of fibrin, and this is probably its main function. Chemically it resembles vitamin E and it can also participate in certain oxidation–reduction reactions. Compounds similar to vitamin K have been detected in bacteria and it may turn out that the vitamin has a separate, quite general role in electron transport. A closely related compound known as ubiquinone or coenzyme Q, which has been detected in animals, plants and microbes and which certainly does function in electron transport, appears to be a vitamin for higher animals and should properly be included with the vitamins of the B group in terms of its biochemical mode of action.

In summary, it appears that one cannot as yet assign a common function to vitamins A, C, D, E and K. They may prove to be co-enzymes of certain differentiated systems, which in the case of A and D at least may be structural components of specific mem-branes, but it is premature to speculate further. The relation be-tween the biochemical role of the individual vitamins and the nature of the disease brought on by deficiency is considerably clearer than in the case of the B group, about the biochemistry of which so much more is known. One may compare, for example, the obvious connection between retinal metabolism and faulty vision (vitamin A), between collagen formation and scurvy (vitamin C), between bone deposition and rickets (vitamin D), or between blood coagulation and haemorrhagia (vitamin K), with the more obscure connection between the decarboxylation of keto acids and beri-beri (vitamin B_1), or between the function of dehydrogenases and inflammation of tongue and mouth (vitamin B_2) or pellagra (nicotinic acid).

5-3 Concluding Remarks

Some of the ways in which hormones and certain vitamins interact with proteins have been discussed. However, the detailed mechanism by which these agents influence enzyme synthesis and enzyme activity remains to be elucidated. The effect of thyroxine in stimulating nuclear activity on the one hand, and the effect of various hormones on adenyl cyclase on the other, may prove to be systems capable of clarifying the molecular basis of hormone action.

5-4 Summary

(a) Hormones

(i) *Enzyme Synthesis.* Hormones are chemical messengers that *initiate* or *maintain* biochemical processes in differentiated target cells.

Thyroxine and ecdysone initiate morphogenetic changes in tadpole and insect larvae respectively. In each case *synthesis* of *specific enzymes* is involved. The same is true of mouse mammary gland cells and of embryonic rat liver cells which respond to hormonal stimulation in culture. Several hormones stimulate protein synthesis in other systems. Induction by hormones may occur through an increase in mRNA synthesis. *Hormonal initiation* cannot be the *primary* cause of differentiation, since the formation of specific *receptor* mechanisms must precede the action of hormones. Moreover certain differentiated cells can be cultured in the absence of hormones.

(ii) *Enzyme Activation*. Insulin and other hormones stimulate transport processes that are enzymic in nature; the mechanism is by *activation* of existing enzymes. Enzymic activation is also responsible for the increase in cyclic AMP that results from the action of certain hormones on specific target cells; cyclic AMP subsequently triggers off further biochemical events.

(iii) *Reaction with Membranes*. The action of hormones, in causing both the *synthesis* and *activation* of enzymes may be correlated with the fact that both processes are linked to the membranes of cells; several hormones have been shown to bind to specific proteins associated with membranes.

(b) Vitamins

Vitamins are dietary factors necessary for the growth of higher animals. The *B group* of vitamins function as coenzymes in general metabolic sequences common to animals, plants and microbes whereas vitamins *A, C, D, E* and *K* act only on certain *differentiated* tissues of higher animals and thus resemble *hormones* in their specificity towards target cells. The mechanism of their action is not yet fully clear but may be coenzymic in nature.

Selected Bibliography

Hormones

General Reading
Bittar, E. E. and N. Bittar (Ed.) (1968). Control of Metabolic Processes. In
   The Biological Basis of Medicine. Academic Press, London–New York,
   Vol. 2, Part 2
Karlson, P. (Ed.) (1965). *Mechanisms of Hormone Action.* Academic Press,
   London–New York
Litvack, G. and D. Kritchevsky (Ed.) (1964). *Action of Hormones on
   Molecular Processes.* John Wiley & Sons, London–New York
Tepperman, J. (1968). *Metabolic and Endocrine Physiology,* 2nd Edition,
   Year Book Medical Publishers, Chicago, Ill.
White, A., P. Handler and E. L. Smith (1968). In *Principles of Biochemistry,*
   4th Edition, McGraw-Hill, London–New York, Chapter 42, p. 917

Protein and RNA Synthesis
Edelman, I. S. and G. M. Fimognari (1969). On the Biochemical Mechanism
   of Action of Aldosterone. *Recent Progress Hormone Research,* **24,** 1
Frieden, E. (1967). Thyroid Hormones and the Biochemistry of Amphibian
   Development. *Recent Progress Hormone Research,* **23,** 139
Goldwasser, E. (1966). Biochemical Control of Erythroid Cell Development.
   Current Topics in Dev. Biol., **1,** 173
Grant, J. K. (1969). Actions of Steroid Hormones at Cellular and Molecular
   Levels. In *Essays in Biochemistry* (Ed. P. N. Campbell and G. D.
   Greville), Academic Press, London–New York, Vol. 5, p. 1
Karlson, P. and C. E. Sekeris (1966). Ecdysone, an Insect Steroid Hormone,
   and Its Mode of Action. *Recent Progress Hormone Research,* **22,** 473
O'Malley, B. W., W. L. McGuire, P. O. Kohler and S. G. Korenman (1969).
   Studies on the Mechanism of Steroid Hormone Regulation of Synthesis
   of Specific Proteins. *Recent Progress Hormone Research,* **25,** 105
Sekeris, C. E. and P. Karlson (1966). Biosynthesis of Catecholamines in
   Insects. *Pharmacol. Rev.,* **18,** 89
Tata, J. R. (1966). Hormones and the Synthesis and Utilization of Ribo-
   nucleic Acids. *Progress Nucleic Acid Research,* **5,** 191
Tata, J. R. (1969). The Action of Thyroid Hormones. *General and Com-
   parative Endocrinology.* Suppl. **2,** 385
Turkington, R. W. (1968). Hormone-Dependent Differentiation of Mam-
   mary Gland *in vitro. Current Topics in Dev. Biol.,* **3,** 199
Wool, I. G., W. S. Stirewalt, K. Kurihara, R. B. Low, P. Bailey and D.
   Oyer (1969). Mode of Action of Insulin in the Regulation of Protein
   Biosynthesis in Muscle. *Recent Progress Hormone Research,* **24,** 139

Enzyme Activation
Christensen, H. N. (1962). *Biological Transport.* W. A. Benjamin, New
   York
Krebs, E. G. and E. H. Fisher (1962). Molecular Properties and Trans-

formations of Glycogen Phosphorylase in Animal Tissues. *Adv. Enzymol.*, **24**, 263

Nahara, H. T. and C. F. Cori (1968). Hormonal Control of Carbohydrate Metabolism in Muscle. In *Carbohydrate Metabolism and its Disorders* (Ed. F. Dickens, P. J. Randle and W. J. Whelan), Academic Press, London–New York, Vol. 1, p. 375

Park, C. R., O. B. Crofford and T. Kono (1968). Mediated (non-active) Transport of Glucose in Mammalian Cells and its Regulation. *J. Gen. Physiol.*, **52**, 296

Robison, G. A., R. W. Butcher and E. W. Sutherland (1968). Cyclic AMP. *Ann. Rev. Biochem.*, **37**, 149

Stein, W. D. (1968). The Transport of Sugars. *Brit. Med. Bull.*, **24**, 146

Sutherland, E. W., I. Oye and R. W. Butcher (1965). The Action of Epinephrine and the Role of the Adenyl Cyclase System in Hormone Action. *Recent Progress Hormone Research*, **21**, 623

Receptors and Membranes

Beermann, W. and U. Clever (1964). Chromosome Puffs. *Scientific American*, **210**, No. 4, p. 50

Clever, U. (1965). The Control of Gene Activity as a Factor of Cell Differentiation in Insect Development. In *Developmental and Metabolic Control Mechanisms and Neoplasia*. The Williams and Wilkins Company, Baltimore, Md., p. 361

Danielli, J. F., J. F. Moran and D. J. Triggle (Ed.) (1970). *Fundamental Concepts in Drug Receptor Interactions*. Academic Press, London–New York

Gorski, J., G. Shyamala and D. Toft (1969). Interrelationships of Nuclear and Cytoplasmic Estrogen Receptors. *Current Topics in Dev. Biol.*, **4**, 149

Gorski, J., D. Toft, G. Shyamala, D. Smith and A. Notides (1969). Hormone Receptors: Studies on the Interaction of Estrogen with the Uterus. *Recent Progress Hormone Research*, **24**, 45

Laufer, H. (1965). Developmental Studies of the Dipteran Salivary Gland, Vol. 3. Relationships Between Chromosomal Puffing and Cellular Function during Development. In *Developmental and Metabolic Mechanisms and Neoplasia*, The Williams and Wilkins Company, Baltimore, Md., p. 237

Peters, R. A. (1956). Hormones and the Cytoskeleton. *Nature*, **177**, 426

Segal, S. J. (1967). Regulatory Action of Estrogenic Hormones. In *Control Mechanisms in Developmental Processes* (Ed. M. Locke), Academic Press, London–New York, p. 264

Tata, J. R. (1967). The Formation and Distribution of Ribosomes During Hormone-Induced Growth and Development. *Biochem. J.*, **104**, 1

Young, F. G. (1962). On Insulin and its Action. *Proc. Roy. Soc. B*, **157**, 1

Vitamins

De Luca, H. F. (1969). Recent Advances in the Metabolism and Function of Vitamin D. *Fed. Proc.*, **28**, 1678

Dingle, J. T. and J. A. Lucy (1965). Vitamin A, Carotenoids and Cell Function. *Biol. Rev.*, **40**, p. 422

Draper, H. H. and A. S. Csallamy (1969). Metabolism and Function of Vitamin E. *Fed. Proc.*, **28**, 1690

Morton, R. A. (Ed.) (1965). *Biochemistry of Quinones*. Academic Press, London–New York

Norman, A. W. (1968). The Mode of Action of Vitamin D. *Biol. Rev.*, **43**, 97

Olson, J. A. (1969). Metabolism and Function of Vitamin A. *Fed. Proc.*, **28**, 1670

Olson, R. E. (1966). Studies on the Mode of Action of Vitamin K. *Adv. Enzyme Regulation*, **4**, 181

Olson, R. E. and P. C. Carpenter (1967). The Regulatory Function of Vitamin E. *Adv. Enzyme Regulation*, **5**, 325

Olson, R. E., R. K. Kipfer and L.-F. Li (1969). Vitamin K-induced Biosynthesis of Prothrombin in the Isolated Perfused Rat Liver. *Adv. Enz. Reg.*, **7**, 83

Papaconstantinou, J. (1967). Metabolic Control of Growth and Differentiation in Vertebrate Embryos. In *Biochemistry of Animal Development* (Ed. R. Weber), Academic Press, London–New York, Vol. 2, p. 58

Pasternak, C. A. and D. B. Thomas (1969). Metabolism of Sulfated Mucopolysaccharides in Vitamin A Deficiency. *Am. J. Clin. Nutrition*, **22**, 986

Robinson, F. A. (1966). *The Vitamin Co-factors of Enzyme Systems*. Pergamon Press, Oxford

Suttie, J. W. (1969). Control of Clotting Factor Biosynthesis by Vitamin K. *Fed. Proc.*, **28**, 1696

White, A., P. Handler and E. L. Smith (1968). *Principles of Biochemistry*, 4th Edition, The Lipid Soluble Vitamins. McGraw-Hill, New York, Chapter 50, p. 1048

CHAPTER 6

Pathological consequences: cancer

Attention was drawn in Section 4-3(*a*) to the observation that the more differentiated certain cells become, the slower they divide. It is also often true that the less differentiated a cell becomes, the more rapidly it divides. The best example of such 'dedifferentiation' is given by the various diseases which share the common property of unlimited growth and are called cancer. Cancer occurs in plants as well as animals; in this chapter cancers of animal origin will be discussed. No attempt will be made to review in this chapter all that is known of the biochemistry of cancer. Instead some specific examples of the alteration of structural and functional proteins in cancer cells will be considered, together with an evaluation of the extent to which such variation underlies the mechanism of carcinogenesis.

6-1 Nature and Aetiology of Cancer

One must distinguish carefully between *rapid* and *uncontrolled* growth of cells. When a part of the liver or one of the two kidneys is surgically removed, the animal does not die. Instead the rest of the liver undergoes rapid cell division until the original size has been restored. In the case of kidney the remaining organ also increases in size until its functional capacity is just about double the normal value. In each case of such organ *regeneration*, the duration of rapid growth is terminated by some mechanism, the biochemistry of which is one of the unsolved areas of differentiation (Section 8-2). In the case of cancer, the cells continue to divide; they can be derived from liver, kidney, muscle, blood or most other tissues of the body, hence the *variety* of different cancers.

Eventually the cells, which may diffuse into the blood stream and spread to other tissues, disrupt some basic physiological process and the animal dies.

The exact origin or aetiology of cancer is unknown. Two possible mechanisms of carcinogenesis will be considered; the first involves genetic change, the second does not. Very many agents have been shown to be in some way responsible for one form of cancer or another. Chemicals, including nicotine, caffeine and various dyes, radiations such as x-rays and the rays accompanying radioactive decay, and viruses have all been implicated; sometimes cancers seem to be quite spontaneous. It is at first sight difficult to incorporate these diverse causes into a coherent mechanism. On the other hand it is a striking fact that several carcinogenic chemicals and radiations have the common property of being highly mutagenic in microbes. That is, they increase by a factor of 1000 or more the rate at which random mutations occur. The mechanism by which mutation is achieved has been shown to be by addition, deletion or alteration of base pairs in DNA. Moreover some bacterial viruses become attached to the chromosome of their host and there is evidence that carcinogenic viruses may behave in a similar manner. An animal cell in which the genes controlling growth have been modified or lost might soon outgrow its neighbours and the phenomenon of cancerous growth would result. A mechanism of carcinogenesis based on such *somatic mutation*, that is a mutation involving only the genes of the cells in which it occurs, is an attractive one at present. It explains, for instance, why virtually any cell-type capable of division can give rise to cancer.

The alternative hypothesis is that genetic change is not involved, but that a lack of growth control is somehow initiated by carcinogens and maintained by cytoplasmic inheritance through subsequent generations. This is not to say that genes are *never* altered. Indeed in certain cancers whole chromosomes are sometimes affected. What is implied is that genetic change is not a *necessary* factor for carcinogenesis. No satisfactory evidence in favour of such a hypothesis has yet emerged, but models of possible mechanisms have been proposed. The particular relevance of cancer to the study of differentiation is that the second hypothesis is exactly that which has come to be accepted as the basis of cellular differentiation (Chapters 7 and 8). Since acceptance of this hypothesis is based on a demonstration that differentiated cells have totipotency and therefore an intact complement of genes (see Section 7-1), it would seem that a similar experiment with cancer cells (such as injecting nuclei of amphibian cancer cells into a normal enucleated egg) might resolve the problem. Such an experiment has been carried out by McKinnell, and the result suggests that in this case, at least, a full complement of genes is retained during carcinogenesis. A somewhat similar experiment with a plant tumour has led to the same result (see Section 7-1(b)).

Even if the mechanism of carcinogenesis in other situations turns out to be by somatic mutation, the fact that cancerous growth of cells is accompanied by altered patterns of metabolism and structure makes the phenomenon phenotypically comparable to that of differentiation, except that the direction of change tends to be from a high degree of differentiation in the cell of origin to a lower one in the ensuing cancer cells. Such a process is not unlike the development of a fertilized egg cell (Section 3-1) in reverse; hence the use of the term 'dedifferentiation'. An indication of an actual similarity between embryonic and cancer cells is discussed under cell-specific antigens below. Because cancer cells are insensitive to restrictions of the environment they are easy to grow in culture, and fruitful biochemical studies have been carried out in this way.

6-2 Properties of Cancer Cells

(a) *Alterations in Metabolism*

(i) *Concentration of Enzymes.* As long ago as 1923 Otto Warburg found that several different types of animal cancer have a high rate of aerobic glycolysis, that is the degradation of glucose only as far as lactic acid, despite the presence of oxygen. The rate of anaerobic glycolysis is also high (Table 6-1). Warburg proposed that these changes are due to a fault in normal respiration and that this is the basic defect of cancer cells. Since then it has been found that several normal types of cells such as retina, placenta and brain also have high rates of glycolysis, whereas in some cancerous ones such as 'minimal-deviation hepatomas' the rate is actually lower (Table 6-1). 'Minimal-deviation hepatomas' are mild forms of liver cancer induced by treating rats with carcinogenic chemicals; the resulting tumours are then implanted into the leg muscles of recipient rats. The tumours proliferate only slowly and appear to differ relatively little, apart from unrestricted growth,

Table 6-1 Glycolytic rate of normal and neoplastic rat tissues[a]

Tissue	Aerobic	Anaerobic
Normal		
Liver	< 1	2
Intestinal mucosa	1	3
Muscle	2	4
Kidney	6	4
Placenta	7	10
Brain	2	14
Embryo (whole)	4	17
Retina	34	66
Tumour		
Minimal-deviation hepatoma 5123-D	< 1	3
Minimal-deviation hepatoma H-35	< 1	2
Flexner–Jobling	19	23
Jensen sarcoma	13	25

[a] Rates are expressed as μmoles lactate released/g tissue/min under oxygen (aerobic) or nitrogen (anaerobic). From Knox (1967) and Weber (1968).

from liver of normal or recipient animals. In other words, these tumours are only slightly dedifferentiated and are therefore useful tools for the study of possible biochemical differences between normal and cancer cells. Individual glycolytic enzymes also are lower rather than higher in minimal-deviation hepatomas compared with liver. In the more malignant tumours, on the other hand, the rate of anaerobic glycolysis and the concentration of key enzymes such as hexokinase, phosphofructokinase and pyruvate kinase, do seem to increase in almost direct proportion to the growth rate (Figure 6-1). These results suggest that high glycolytic rate may be a prerequisite for rapid growth and thus a consequence rather than a cause of malignancy.

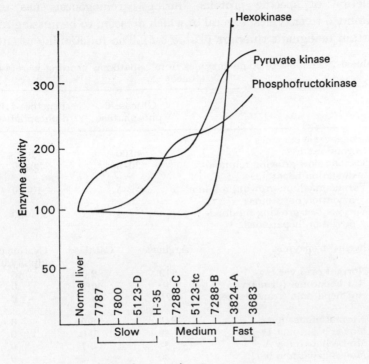

Figure 6-1 Glycolytic enzymes (see Figure 4-1) of some minimal-deviation hepatomas of differing growth rate. Enzyme activities are expressed as a percentage of that of normal liver. From Weber (1968)

Several chemical carcinogens, such as the dye 'butter-yellow' (Figure 6-2) bind to specific cellular proteins during the period of

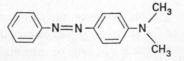

Figure 6-2 4-Dimethylaminoazobenzene ('butter-yellow'), a potent carcinogen

5

induction ('precancerous stage'). By the time the cells become malignant, these proteins can no longer be detected. The physiological function of such proteins is not clear, but the idea of a deletion of specific proteins during carcinogenesis has been extended to enzymes, several of which do seem to be missing from certain malignant tumours (Table 6-2). The functioning of entire

Table 6-2 The loss of some enzymes from hepatomas growing in rats and mice[a]

	Glucose-6-phosphatase	Fructose-1:6-diphosphatase
Gluconeogenic enzymes		
Normal rat liver	100	100
Various slow-growing minimal-deviation hepatomas	20–60	30–60
Various medium-growing minimal-deviation hepatomas	< 2–5	10–20
Various fast-growing minimal-deviation hepatomas	< 1	< 1–30

Other liver enzymes	Arginase	Catalase	Cystine de-sulphhydrase
Normal rat liver	213	2·0	45
Rat hepatoma (primary)	40	0·1	0
Rat hepatoma (transplanted)	21	0	0
Normal mouse liver	246	8·0	6
Mouse hepatoma 1	40	0·6	0
Mouse hepatoma A	34	0·1	0
Mouse hepatoma 98/15	30	0·1	0

Arginase: arginine → urea + ornithine (see Figure 3-8)
Catalase: $H_2O_2 \rightarrow O_2 + H_2O$
Cystine desulphhydrase:
     cystine → pyruvate + NH_3 + H_2S

Activities of gluconeogenic enzymes (see Figure 4-1) are expressed relative to that of rat liver. Other activities are in arbitrary units
[a] See Figure 6-1 for details of the minimal-deviation hepatomas. From Greenstein (1954) and Weber (1968)

pathways, such as gluconeogenesis, may be reduced. Cancer cells then rely on the host for their supply of essential nutrients (see Section 6-2(b)). However this is not true of minimal-deviation hepatomas, and hence enzyme deletion, like raised glycolytic rate, appears to be an outcome, not a cause of cancer.

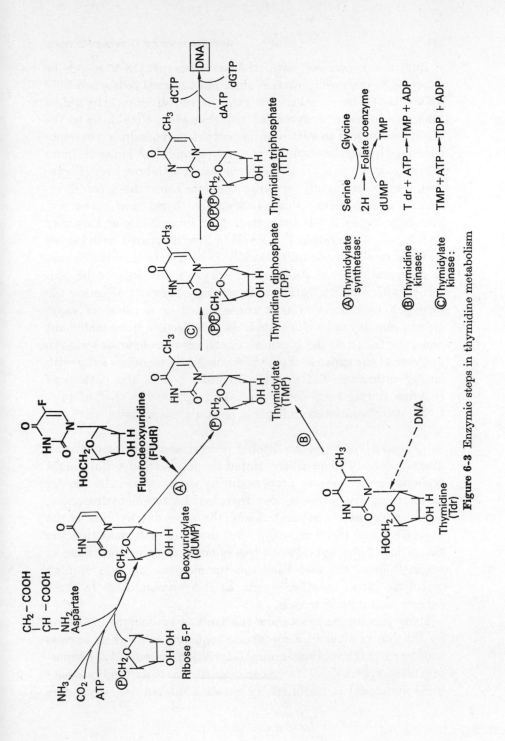

Figure 6-3 Enzymic steps in thymidine metabolism

Enzymes concerned with the synthesis of DNA might be expected to increase in cancer, since most normal cells when fully differentiated have rather low rates of cell division. Thymidine kinase (a 'salvage' enzyme for any thymidine circulating in the blood stream and therefore an important enzyme during carcinogenesis), thymidylate synthetase and thymidylate kinase (Figure 6-3) all increase in concentration compared with host liver (Figure 6-4). The most rapidly growing tumours show the greatest increase, though even minimal-deviation hepatomas have significantly higher levels than their host liver; however this may merely reflect the rather low levels in liver compared with tissues which normally divide quite rapidly, such as intestinal epithelium or regenerating liver. Again these enzymes are likely to be concerned with rapidity, rather than extent, of growth. Whether the increased concentration is a consequence or a cause of rapid growth remains to be elucidated. As cells become more malignant and grow faster, so the synthesis of cell-specific enzymes seems to decrease at the expense of enzymes concerned predominantly with energy utilization and growth. In other words, the pattern of enzymes during such dedifferentiation approaches that of relatively undifferentiated embryonic or adult 'stem' cells.

(ii) *Properties of Enzymes*. Altered properties of existing enzymes, characteristic of some differentiated tissues (Section 4-2(*b*)) might account for the reverse process during carcinogensis, but so far no startling changes have been revealed. Lactate dehydrogenase, for example, can be separated into the same five hybrids as the normal enzyme (Section 4-2(*b*)). But in rather malignant tumours the ratios of the hybrids are less characteristic of the tissue of origin than of the fact that the tumours are capable of high glycolytic rates. In other words LDH-5 predominates in these cancers as it does in muscle.

Many tumour enzymes show the same sensitivity to inhibition by the end product of a metabolic sequence as do their normal counterparts. Thus in one minimal-deviation tumour (5123), aspartate transcarbamylase (the first enzyme specific to pyrimidine nucleotide synthesis) is inhibited by cytidine nucleotides, thymidine

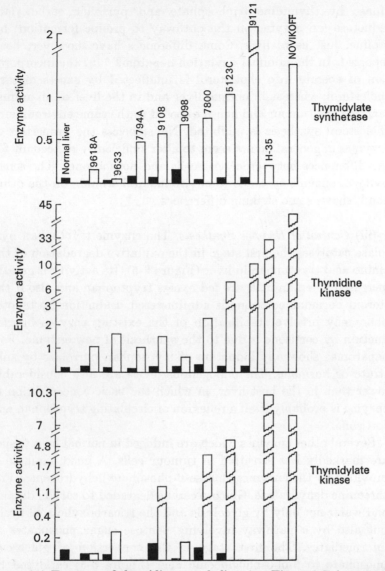

Figure 6-4 Enzymes of thymidine metabolism (see Figure 6-3) in normal rat liver and in various liver tumours. 9618A–9121 are minimal-deviation hepatomas grown in rats; the enzyme levels in the corresponding host livers are also shown in black. H-35 (another minimal-deviation hepatoma) and Novikoff (a rather malignant hepatoma) were grown in culture. Enzyme activities are expressed as μmoles product formed/g protein/hr. From Sneider, Potter and Morris (1969) *Cancer Res.*, **29**, 40

kinase by thymidine triphosphate and pyrroline carboxylate reductase (an enzyme on the pathway to proline formation) by proline, just as in liver. Some differences have, however, been reported. In the minimal-deviation hepatoma 7787 the incorporation of acetate into cholesterol is unaffected by excess dietary cholesterol, whereas in normal liver and in the liver of an animal carrying the tumour and hence exposed to the same environment, cholesterol synthesis is inhibited. Nevertheless the properties of enzymes in general do not seem to alter sufficiently to account for the differences between cancer cells and normal ones. The sensitivity of enzyme *synthesis* to environmental stimuli, on the other hand, shows some striking differences.

(iii) *Control of Enzyme Synthesis.* The enzyme tryptophan pyrrolase catalyses the first stage in the oxidative degradation of the amino acid tryptophan in liver (Figure 6-5). Its activity is greatly increased when animals are fed excess tryptophan and also if the steroid hormone cortisone is administered. Induction by tryptophan may involve stabilization of the existing enzyme but induction by cortisone is due to the synthesis of new enzyme. Few hepatomas show any induction of tryptophan pyrrolase by substrate or hormone. The endogenous levels are also considerably lower than in the host liver, in which the basic concentration of enzyme is probably itself a reflexion of circulating tryptophan and cortisone.

Several other enzymes which are induced in normal host tissues are markedly less sensitive in tumour cells. A good example is provided by the enzymes glucose-6-phosphate dehydrogenase and threonine dehydratase. Glucose can be degraded to carbon dioxide and water not only by glycolysis and the tricarboxylic acid cycle, but also by a pathway involving pentose sugar phosphates as intermediates. The first stage is the conversion of glucose-6-phosphate to 6-phosphogluconic acid (Figure 6-5) catalysed by glucose-6-phosphate dehydrogenase. Diets high in carbohydrate (60 per cent carbohydrate, 30 per cent protein) induce the enzyme. Threonine dehydratase catalyses the first reaction in the degradation of threonine (Figure 6-5) and is induced by diets rich in

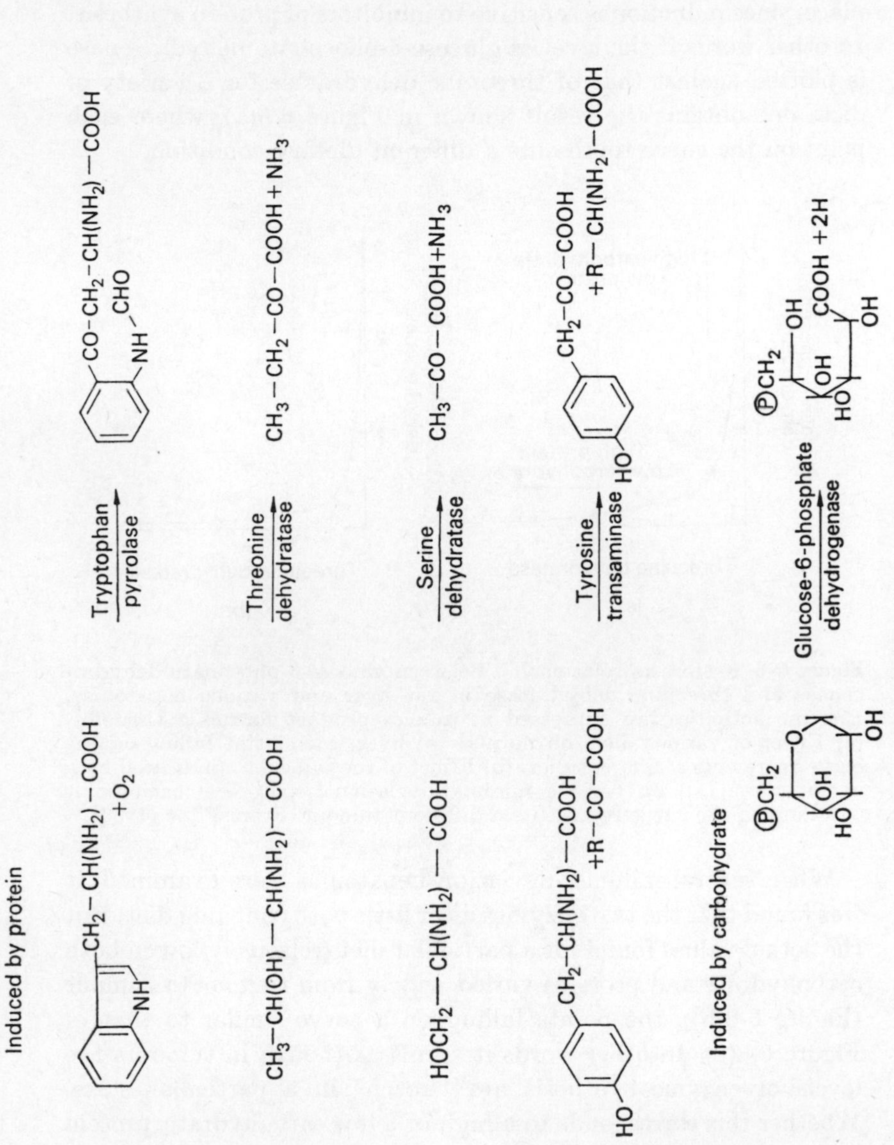

Induced by protein

Tryptophan:

$CH_2-CH(NH_2)-COOH + O_2$ (indole) $\xrightarrow{\text{Tryptophan pyrrolase}}$ $CO-CH_2-CH(NH_2)-COOH$ (with $NH-CHO$)

$CH_3-CH(OH)-CH(NH_2)-COOH$ $\xrightarrow{\text{Threonine dehydratase}}$ $CH_3-CH_2-CO-COOH + NH_3$

$HOCH_2-CH(NH_2)-COOH$ $\xrightarrow{\text{Serine dehydratase}}$ $CH_3-CO-COOH + NH_3$

Tyrosine:

$CH_2-CH(NH_2)-COOH + R-CO-COOH$ $\xrightarrow{\text{Tyrosine transaminase}}$ $CH_2-CO-COOH + R-CH(NH_2)-COOH$

Induced by carbohydrate

Glucose-6-phosphate $\xrightarrow{\text{Glucose-6-phosphate dehydrogenase}}$ $COOH + 2H$

Figure 6-5 Some inducible enzymes of rat liver

protein. In each case synthesis of new enzyme appears to take place, since induction is sensitive to inhibitors of protein synthesis. In other words if the level of glucose-6-phosphate dehydrogenase is plotted against that of threonine dehydratase for a variety of diets one obtains the result shown in Figure 6-6(a), where each point on the curve represents a different dietary condition.

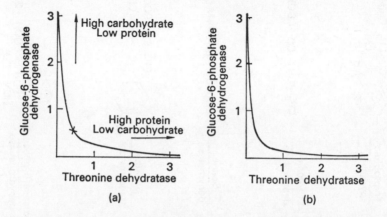

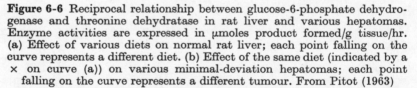

Figure 6-6 Reciprocal relationship between glucose-6-phosphate dehydrogenase and threonine dehydratase in rat liver and various hepatomas. Enzyme activities are expressed in μmoles product formed/g tissue/hr. (a) Effect of various diets on normal rat liver; each point falling on the curve represents a different diet. (b) Effect of the same diet (indicated by a × on curve (a)) on various minimal-deviation hepatomas; each point falling on the curve represents a different tumour. From Pitot (1963)

When several minimal-deviation hepatomas were examined, it was found that the two enzymes alter little with changing diet. But the actual values found for a particular diet (relatively low in both carbohydrate and protein) varied widely from tumour to tumour (Figure 6-6(b)), the points falling on a curve similar to that of Figure 6-8(a). In other words it seems as though in tumours the levels of enzyme synthesis are 'frozen' in a particular state. Whether this corresponds to a high or a low carbohydrate:protein ratio presumably depends on the internal milieu of the cancer cells and is independent of external conditions.

Essentially similar results have been obtained with two other enzymes (Figure 6-5), involved in amino acid, degradation which are normally induced by diets rich in protein. Thus increasing a diet from 12 to 60 per cent protein has little effect on tyrosine trans-aminase or serine dehydratase in any of nine minimal-deviation hepatomas studies, whereas the level in the corresponding host livers increased some 10 to 20-fold.

A general insensitivity to environmental control may prove to be an important factor in explaining the unlimited proliferation of cancer cells. It is, however, more likely to be a prerequisite than a cause of cancer. Lack of response to external variation may be due to altered surface membranes, and evidence on this point will now be examined.

(b) Alterations in Structure

(i) *Permeability Changes.* The direct examination of proteins and other components of membranes has proved difficult so far. This is because of the problem of isolating a pure surface membrane free of other cell debris. On the other hand the consequences of altered membranes are easier to investigate, as illustrated by the discussion on glucose transport in Section 5-1(c). One result that has been established is that the transport of amino acids into certain cancer cells, albeit rather malignant ones, is very much greater than into normal cells, the tumours achieving an internal concentration some 10-fold in excess of that outside. This is reflected in a decreased level of some amino acids in the blood stream of such tumour-bearing animals. The ability to 'mop up' amino acids and other nutrients by increased permeability or subsequent metabolism, and to maintain such compounds at high internal concentration, may partly explain both the unresponsive-ness and the rapid growth rate of certain tumours.

(ii) *Cell-specific Antigens.* Another consequence of altered mem-branes might be the presence of altered surface antigens, since it was seen in Section 4-3(b) that differentiated cells contain organ-specific antigens. Cancers induced in inbred strains of mice by carcinogenic chemicals or viruses have indeed revealed the presence of specific antigens that are different from any host antigen.

Attempts to immunize mice against respective carcinogens have had some success. This has naturally stimulated efforts to find cancer-specific antigens in humans and an interesting result has recently emerged.

It appears that cancer cells from the human colon contain antigens which are different from those of neighbouring undiseased colonic tissue. Colonic cancers from many different patients have the same antigens. A very intriguing observation is that while these antigens cannot be detected in any tissue of normal *adults*, they are present in *embryonic* and *foetal* organs connected with the digestive system, such as differentiating intestine, liver or pancreas, but are absent from other developing organs. In other words, it appears that the cell-specific production of cancer antigens that occurs during normal foetal development is somehow halted after birth and the antigens are lost or masked; it continues or re-commences only in individuals who develop intestinal cancer. Antigenic masking has been observed in other systems. Whether such antigens are a cause, a prerequisite or a consequence of cancer is not known. The fact that they are protein–carbohydrate complexes situated on the surface membrane of cells (like the mouse cancer antigens mentioned earlier) may be important in relation to the failure of many cancer cells to exhibit 'contact inhibition' (see below). The cessation of their synthesis during normal development fits in with some recent results obtained by the technique of cellular fusion.

A line of low malignancy (as defined by the incidence of tumours following injection of the cells into mice) called A9 was fused with different lines of high malignancy such as Ehrlich ascites cells. In each case, the resulting hybrids were found to be of low malignancy; in other words, malignancy can be suppressed (Figure 6-7). It should be pointed out that this result is in contrast to the conclusions reached by other workers. In the hybrids, expression of the antigenicity associated with A9 (which is stronger than that of the malignant parent) is also suppressed. Suppression of malignancy may be associated with a gene product of the non-malignant parent, since one in 10^6–10^7 of hybrid cells appears to regain malignancy and this is accompanied by a loss of several chromo-

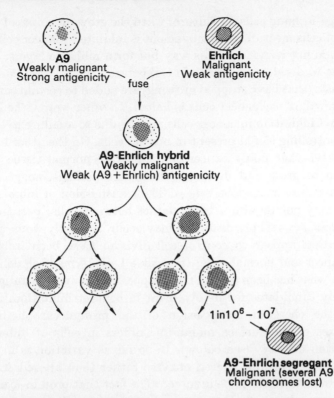

Figure 6-7 Suppression of malignancy by cell fusion (see Figure 7-5). Fusion of A9 cells with Ehrlich ascites cells results in hybrid cells of low malignancy. Hybrids containing one nucleus from each parent undergo subsequent cell divisions. During these divisions a few 'segregants' appear in which malignancy has been regained; at the same time several A9 chromosomes are lost. Drawn to illustrate the results of Harris, Miller, Klein, Worst and Tachibana (1969)

somes known to be derived from A9 cells. Whether the malignant parent lacks such a 'suppressor gene' or whether one is present but inactive, is not yet known. When this point becomes clear, it should be possible to distinguish between the mutational and non-mutational hypothesis of carcinogenesis, at least so far as malignancy in this system is concerned.

(iii) *Contact Inhibition.* Contact inhibition is the term used to describe the behaviour of some normal cells such as fibroblasts (cartilage-forming cells) in culture: when the growing points of two layers of cells meet, further movement is inhibited. Cancer cells in general do not respond in this way but form multiple layers, and continue to divide regardless of neighbouring cells. When normal fibroblasts (that have stopped growing) are added to certain cancer cells, growth of the cancer cells is halted. In other words, the lack of contact inhibition in cancer cells may be due to a failure to emit some controlling signal present in normal cells. On the other hand it is known that many cancer cells infiltrate normal tissue and continue to grow and divide in such surroundings, suggesting rather a failure in responsiveness. The transmission of inhibitory signals may not require actual contact between cells; passage of substances across short distances may occur. Recently, some substances that appear to control cell division have been isolated from cancer and normal cells; they have been termed 'chalones'.

From what has been said so far, it appears that specific antigens are likely candidates for involvement in contact inhibition. Certainly the chemical composition of the protein:carbohydrate complexes of the surface membrane differs in cells of differing 'saturation density' (see below). In so far as variation is in the carbohydrate portion, altered enzyme rather than altered structural protein appears to be involved. The fact that protein–carbohydrate complexes, also known as glycoproteins, may actually be extruded from certain cultured cells is in accord with the possibility that contact inhibition operates extracellularly over short distances.

The importance of contact inhibition to growth control is well illustrated in the case of some virally-induced cancer cells. These cells, which form tumours when injected into mice, are killed in culture by fluorodeoxyuridine, a drug which inhibits the *de novo* synthesis of thymidylate (Figure 6-3). This prevents DNA synthesis and hence dividing cells are killed; non-growing cells are not affected. Cells resistant to fluorodeoxyuridine have been isolated. It is found that they have a 'saturation density' (cells per unit area at which growth stops) considerably lower than sensitive cells (Figure 6-8). The original insensitivity of resistant cells to fluorodeoxyuridine is therefore due to the fact that they had

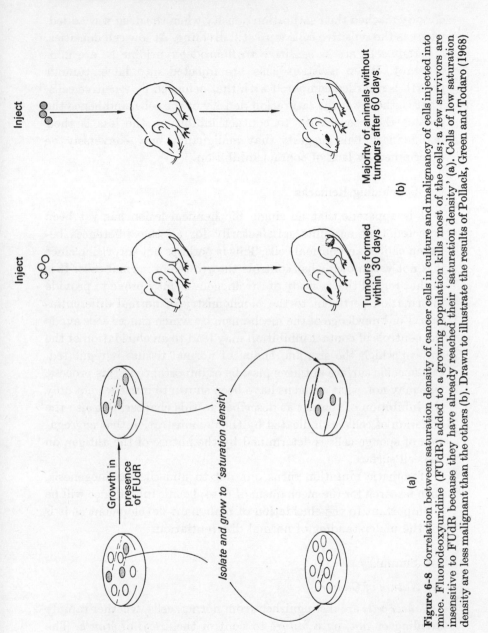

Figure 6-8 Correlation between saturation density of cancer cells in culture and malignancy of cells injected into mice. Fluorodeoxyuridine (FUdR) added to a growing population kills most of the cells; a few survivors are insensitive to FUdR because they have already reached their 'saturation density' (a). Cells of low saturation density are less malignant than the others (b). Drawn to illustrate the results of Pollack, Green and Todaro (1968)

already reached their saturation density when the drug was added, whereas the sensitive cells were still dividing. At low cell densities, resistant cells are as sensitive to fluorodeoxyuridine as are non-resistant. When resistant cells are injected into mice, tumour growth is retarded compared with that achieved by sensitive cells. That is, the lower the saturation density of the cells, and hence the greater their sensitivity to contact inhibition, the less is their malignancy. This suggests that malignancy may conversely be determined by lack of contact inhibition.

6-3 Concluding Remarks

It is apparent that no single biochemical lesion has yet been detected that accounts satisfactorily for all the differences between cancer and normal cells. This is perhaps not surprising since it is not yet known what controls the growth of normal cells. It is in that regard that a study of carcinogenesis may prove to provide information pertinent to the biochemistry of normal differentiation. For knowledge of the mechanism by which cancer cells evade the controls of contact inhibition may lead to an elucidation of the way in which the size and shape of normal tissues is regulated. Cell-specific surface antigens may be of importance in this process. One may note that antigens have been shown to influence not only the inhibition of contact as described in this chapter, but also the adhesion of cells, as indicated by the demonstration that aggregation of sponge cells is determined by the nature of the antigen on the cell surface.

If somatic mutation turns out not to underlie carcinogenesis, then a search for the mechanism of cytoplasmic inheritance will be as important to the elucidation of malignant development as it is for the understanding of normal differentiation.

6-4 Summary

(a) Nature of Cancer

Cancer cells are distinguished from normal cells, whether rapidly dividing or not, by a *failure* to control the *extent* of *growth*. The defect, which is induced by chemicals, radiations or viruses, may be due to *somatic mutation*, or it may arise from faulty propagation of *cytoplasmic factors*.

(b) *Alterations in Metabolism*

(i) *Concentration of Enzymes.* The rate of *aerobic* and *anaerobic glycolysis* is higher in rather malignant (that is rapidly growing) tumours than in most normal cells, but this is not true of 'minimal-deviation hepatomas' (slow-growing liver cancers). The concentration of key *glycolytic* enzymes (hexokinase, phosphofructokinase and pyruvate kinase) is *proportional* to the *growth rate* of certain tumours. The concentration of enzymes concerned with *DNA synthesis* (thymidine kinase and thymidylate kinase) also increases with growth rate. Hence increased *glycolysis* and *DNA synthesis* are likely to be factors concerned with *rapidity* rather than *extent* of growth.

Certain enzymes not essential to *energy metabolism* or *cell division* are *lost* from some malignant tumours. In general, the more *dedifferentiated* a tumour becomes in this respect, the faster it grows.

(ii) *Properties of Enzymes.* The *properties* of *tumour* enzymes as regards sensitivity to metabolic control are generally *similar* to those of *normal* enzymes.

(iii) *Control of Enzyme Synthesis.* Tumours, including minimal-deviation hepatomas, are generally *unresponsive* to agents such as amino acids that induce the *synthesis* of *enzymes* in normal liver. *Insensitivity* to the *environment* appears to be a general property of *cancer cells* and may be a prerequisite for unlimited growth.

(c) *Alterations in Structure*

(i) *Cell-specific Antigens.* Cancer cells contain *specific surface antigens*, distinct from the antigens of normal cells. Antigens of *colonic cancer* cells from adult humans are found in digestive cells of normal human *embryos*; they disappear normally after birth. *Suppression* of cancer antigens, and of *malignancy* itself, has been observed in *hybrid* cells obtained by fusion of malignant with non-malignant cells.

(ii) *Contact Inhibition.* The *movement* of *cancer cells* in culture, unlike that of normal cells, is not inhibited by *cellular contact*; hence growth and cell division proceed further. *Insensitivity* to *contact inhibition* is proportional to the *malignancy* of certain cancer cells.

No single biochemical alteration has yet emerged that satisfactorily explains the cause of cancer.

Selected Bibliography

General Reading

Berenblum, I. (1964). The Nature of Tumour Growth. In *General Pathology*, 3rd Edition (Ed. H. W. Florey), Lloyd-Luke, London, p. 528

Bittar, E. E. and N. Bittar (Ed.) (1969). The Cancer Cell. In *The Biological Basis of Medicine*, Academic Press, London–New York, Vol. 5, Part 4

Emmelot, P. E. and O. Muhlbock (1964). *Cellular Control Mechanisms and Cancer.* Elsevier Publishing Company, Barking, Essex

Greenstein, J. P. (1954). *Biochemistry of Cancer.* Academic Press, London–New York

Carcinogenesis

Dulbecco, R. (1967). The Induction of Cancer by Viruses. *Scientific American*, **216**, No. 4, p. 28

Gelboin, H. V. (1969). Effect of Carcinogens on Gene Action. In *Exploitable Molecular Mechanisms and Neoplasia.* The Williams and Wilkins Company, Baltimore, Md., p. 285

McKinnell, R. G., B. A. Deggins and D. D. Labat (1969). Transplantation of Pluripotential Nuclei from Triploid Frog Tumors. *Science*, **165**, 394

Miller, E. C. and J. A. Miller (1966). Mechanisms of Chemical Carcinogenesis: Nature of Proximate Carcinogenesis and Interactions with Macromolecules. *Pharmacol. Rev.*, **18**, 805

Pitot, H. C. and C. Heidelberger (1963). Metabolic Regulatory Circuits and Carcinogenesis. *Cancer Research*, **23**, 1694

Potter, V. (1962). Enzyme Studies on the Deletion Hypothesis of Carcinogenesis. In *Molecular Basis of Neoplasia.* University Texas Press, Austin, Texas, p. 367

Rapp, F. and J. L. Melnick (1966). The Footprints of Tumor Viruses. *Scientific American*, **214**, No. 3, p. 34

Roe, F. J. C. (1969). Mechanisms of Carcinogenesis. In *The Biological Basis of Medicine*, ed. E. E. Bittar and N. Bittar, Academic Press, New York–London, Vol. 5, p. 487,

Enzymes and Metabolism

Knox, W. E. (1967). The Enzymic Pattern of Neoplastic Tissue. *Adv. Cancer Res.*, **10**, 117

Pitot, H. C. (1963). Some Biochemical Essentials of Malignancy. *Cancer Res.*, **23**, 1474

Potter, V. R., M. Watanabe, H. C. Pitot and H. P. Morris (1969). Systematic Oscillations in Metabolic Activity in Rat Liver and Hepatomas. Survey of Normal Diploid and other Hepatoma Lines. *Cancer Res.*, **29**, 55

Weber, G. (1968). Carbohydrate Metabolism in Cancer Cells and The Molecular Correlation Concept. *Naturwissenschaften*, **9**, 418

Wenner, C. E. (1967). Progress in Tumor Enzymology. *Adv. Enzymol.*, **29**, 321

Surface Membranes and Antigens

Ambrose, E. J. (1967). Possible Mechanisms of the Transfer of Information between Small Groups of Cells. In *Cell Differentiation, CIBA Foundation Symposium* (Ed. A. V. S. de Reuck and J. Knight), J. & A. Churchill, London, p. 101

Black, P. H. (1968). The Oncogenic DNA Viruses: A Review of *In Vitro* Transformation Studies. *Ann. Rev. Microb.*, **22**, 391

Burger, M. M. (1969). A difference in the Architecture of the Surface Membrane of Normal and Virally Transformed Cells. *Proc. Nat. Acad. Sci.*, **62**, 994 (and 1074)

Curtis, A. S. G. (1967). *The Cell Surface: Its Molecular Role in Morphogenesis*. Logos Press, Academic Press, London–New York

Day, E. D. (1965). *The Immunology of Cancer*, Charles C. Thomas, Springfield, Ill.

Gold, P., M. Gold and S. O. Freedman (1968). Cellular Location of Carcinoembryonic Antigens of the Human Digestive System. *Cancer Res.*, **28**, 1331

Green, H. N., H. M. Anthony, R. W. Baldwin and J. W. Westrop (1967). *An Immunological Approach to Cancer*. Butterworths, London

Green, H. and G. J. Todaro (1967). *The Development of the Transformed State in Mammalian Cells Infected with Oncogenic Viruses in Carcinogenesis*. Williams & Wilkins Company, Baltimore, Md., p. 559

Harris, H., O. J. Miller, G. Klein, P. Worst and T. Tachibana (1969). Suppression of Malignancy by Cell Fusion. *Nature*, **223**, 363

Humphreys, T. (1967). The Cell Surface and Specific Cell Aggregation. In *The Specificity of Cell Surfaces* (Ed. B. D. Davis and L. Warren), Prentice-Hall, Englewood Cliffs, New Jersey, p. 195

Johnstone, R. M. and P. G. Scholefield (1965). Amino Acid Transport in Tumor Cells. *Adv. Cancer Res.*, **9**, 144

Klein, G. (1967). Tumor Antigens. In *The Specificity of Cell Surfaces* (Ed. B. D. Davis and L. Warren), Prentice-Hall, Englewood Cliffs, New Jersey, p. 165

Law, L. W. (1969). Studies of the Significance of Tumor Antigens in Induction and Repression of Neoplastic Diseases. *Cancer Res.*, **29**, 1

Pollack, R. E., H. Green and G. J. Todaro (1968). Growth Control in Cultured Cells: Selection of Sublines with Increased Sensitivity to Contact Inhibition and Decreased Tumor-Producing Ability. *Proc. Nat. Acad. Sci.*, **60**, 126

Prehn, R. T. (1967). *Tumor Antigens in Immunity Cancer and Chemotherapy* (Ed. E. Mihich), Academic Press, London–New York, p. 265

Stoker, M. (1968). Contact and Short Range Interactions Affecting Growth of Animal Cells in Culture. *Current Topics in Dev. Biol.*, **2**, 108

Chalones

Bullough, W. S. (1968). The Control of Tissue Growth. In *The Biological Basis of Medicine*, (Ed. E. E. Bittar and N. Bittar), Academic Press, London–New York, Vol. 1, p. 311

Film on Contact Inhibition

Interference Microscopy of Normal and Cancer Cells. Chester Beatty Research Institute, London.

CHAPTER 7

Mechanism: facts

An attempt was made in the last six chapters to see how far it is possible to account for differentiation in terms of specific enzymes and structural proteins. One may conclude that variations in the type and rate of protein synthesis can account quite satisfactorily for many of the properties of differentiated cells and tissues. The next question that must be asked is by what mechanism the differences arise. In other words, how is protein synthesis controlled so as to produce specific proteins, at defined rates, in discrete tissues. Basically two possibilities exist: one is by variation of genetic expression, the other by alteration of genetic content. It has now become clear that alteration of genetic content is *not* the mechanism. The experiments by which this has been established will be considered in this chapter along with some work which shows that the *stimulus* for certain developmental events is located in the cytoplasm, not the nucleus, of cells.

7-1 Totipotency of Differentiated Cells

Since the structure of proteins is coded by the sequence of bases in the DNA of the corresponding genes, an obvious way of achieving variations in different cells is by alteration of the DNA in the way that species variation is due to altered genomes. Unfortunately this hypothesis cannot be tested directly since the methods are not yet available for analysing the sequence of nucleotides in specific genes. Instead one must turn to the means by which the outcome of altered genes may be revealed. The most obvious consequence would be the inability of a cell once differentiated to turn back into a previous, less differentiated one. Indeed this is exactly how 'commitment' was defined in Chapter 1. But it was cautiously added that there were certain exceptions to this rule.

(a) Microbes

The first example of 'totipotency' is the return of microbial spores to vegetative cells (Figure 1-1), or that of slime mould spores to single myxamoebae (Figure 2-6). In each case the return mechanism, germination, is different from the forward one, so that the definition of commitment actually holds good. What is important is that spores clearly retain the genes necessary for vegetative life and *vice versa*.

(b) Plants

Another example of the retention of genes during differentiation is provided by the common gardening practice of planting cuttings of plant organs such as stems and allowing them to root, though in this case it can be argued that some primary or embryonic cells have persisted in the cutting. Objections can also be raised against experiments in which single cells of a tobacco plant tumour (see Section 6-1) are grafted onto healthy plants; seeds isolated from the graft give rise to intact, healthy, tobacco plants.

The following experiments are more conclusive. Pieces of carrot root phloem, free of cambium, are cultured in a nutrient medium in such a way that free phloem cells are formed; if these are now exposed to coconut milk (a 'natural' medium for germination) containing certain additional hormones, the phloem cells begin to 'redifferentiate' and to form roots; this is followed by the appearance of shoots, and eventually by the development of an intact carrot plant (Figure 7-1). In other words, adult cells have been induced to behave like embryonic cells by exposing them to the nutritional environment of embryonic cells. The concentration of the respective hormones is crucial. At one ratio of kinin to auxin, for example, phloem cells continue to differentiate as phloem. At another, 'redifferentiation' into a developing embryo takes place. A similar situation was described in the case of cultured callus tissue (Section 2-4(d)) in which the relative concentration of hormones determines whether development into phloem or xylem occurs. The presence of hormones is clearly a very important part of the commitment process in plants.

A similar experiment has been performed with tobacco plants.

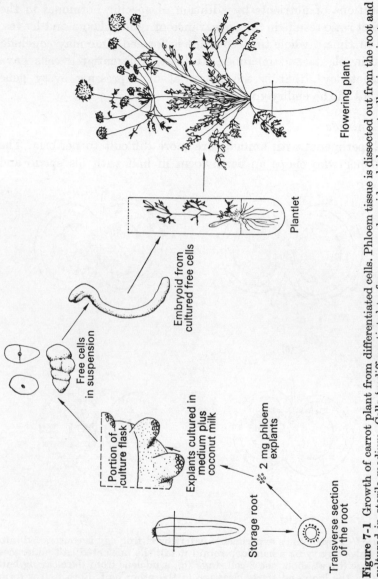

Figure 7-1 Growth of carrot plant from differentiated cells. Phloem tissue is dissected out from the root and cultured in sterile medium. Cells 'redifferentiate' to form an embryoid, which eventually gives rise to an intact carrot plant. From Steward (1964)

Flowering plant

Plantlet

Embryoid from cultured free cells

Free cells in suspension

Portion of culture flask

Explants cultured in medium plus coconut milk

2 mg phloem explants

Storage root

Transverse section of the root

Single cells are isolated from the pith callus of adult stem sand cultured in a chemically defined nutrient medium. Subsequent alterations of nutrients by addition of specific hormones in the correct ratio results in the appearance of roots, shoots and leaves, and in time a whole tobacco plant is formed. One may conclude that in plants and microbes, at least, differentiated cells have 'totipotency', that is the potential of expressing every gene present in the embryonic cell.

(c) *Animals*

Experiments with animals are more difficult to perform. The gardener who chops an earthworm in half with his spade and

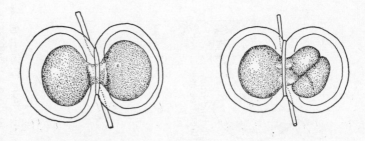

(a)   (b)

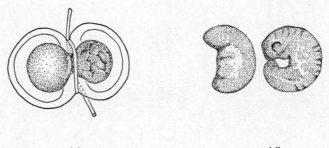

(c)   (d)

Figure 7-2 Spemann's experiment. A fertilized frog egg is constricted into two halves by tying a hair loop round it (a); the nucleated half undergoes cleavage (b); at about the 8-cell stage (c), a nucleus from the cleaving half slips by the loop and starts cleavage in the other half. Both halves form complete embryos, though the one on the left lags behind the other one (d).
From Spemann (1938)

watches the two parts wriggling away deceives himself if he thinks that both parts eventually regenerate a fully-grown worm. The head half, it is true, can be made to regenerate a tail part, but since it contains mesenchyme and other 'primitive' cells from which tail cells are made, this experiment is inconclusive.

Another experimental approach is that of Spemann (1914) who took a frog egg and tied a hair loop round it so that the nucleus was separated from half the egg (Figure 7-2). The nucleated half underwent cleavage and normal development; the enucleated half stayed dormant until a nucleus from the adjoining half at the 8-cell stage slipped by the loop; cleavage then began and both halves developed into normal embryos, though the half that was originally enucleated lagged somewhat behind. Although no visible signs of differentiation are apparent at the 8-cell stage, it is likely that regional commitment to specialization has already occurred. In other words, nuclei from partially differentiated cells retain their original complement of genes. More important, perhaps, is the fact that this experiment showed that nuclei can be successfully transplanted into enucleated eggs.

Gurdon has refined nuclear transplantation to such an extent that he has been able to produce adult frogs from fully differentiated intestinal nuclei (Figure 7-3). First the nucleus of a frog egg is destroyed by surgical removal with a needle or by inactivation with ultraviolet light. Into such a recipient egg (which will not develop on its own) is introduced a nucleus from the intestinal epithelium of a feeding tadpole; nuclei are prepared by separation of individual cells, followed by breakage of their outer membrane by suction through a pipette. Any cytoplasm that is injected along with the nucleus is insignificant; in any case, injection of enucleated cytoplasm is without effect. Not all the eggs develop; sometimes as few as 1–2 per cent, sometimes as many as 30 per cent grow into normal tadpoles, according to the degree of differentiation of the cells used to provide the nuclei. It cannot be argued that the percentage of successes reflects the number of situations in which the recipient nucleus has not been completely destroyed since enucleation is not nearly such an inefficient process; moreover the tadpoles which develop have the genetic

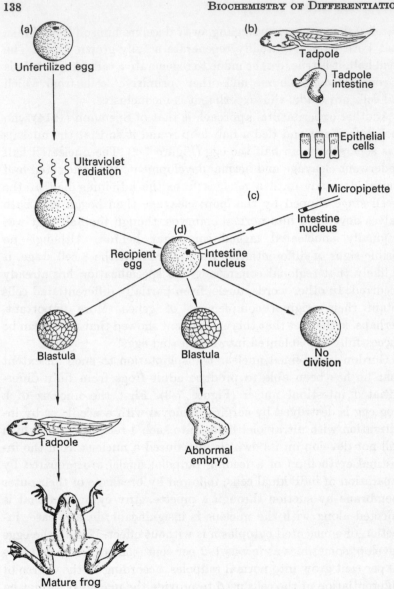

Figure 7-3 Nuclear transplantation in frog eggs. The nucleus of the recipient egg is destroyed by ultraviolet radiation (a). Intestinal epithelial cells are isolated from a tadpole (b) and single cells ruptured by suction through a micropipette (c). A nucleus is then injected into the recipient egg (d). Not all eggs develop, but a significant number form intact frogs.
From Gurdon (1968b)

characteristics of the donor, not the recipient, animal. This is shown by injecting nuclei from the intestine of a pigmented tadpole into non-pigmented eggs; pigmented tadpoles only are formed. Similar transplantation experiments have been carried out by other workers (e.g. Section 6-1).

A quite different approach is to compare the chemical composition of the DNA of different organs. Using the hybridization technique (Section 8-1(b)) it has been shown that DNA preparations from embryonic mouse and from adult mouse brain, kidney, thymus, spleen or liver are indistinguishable from each other; so, incidentally, is DNA from a mouse cancer cell (Section 6-1). However the sensitivity of the hybridization procedure is insufficient to detect differences which would arise from the loss of only a few genes. Absolute intactness of genes during differentiation cannot therefore be deduced from this experiment.

Gurdon's experiments are not subject to such a proviso. Thus in frogs as well as plants, differentiated cells have totipotency, and one may assume that this is true of all living organisms.

(d) Intactness of Genetic Material

Totipotency has been interpreted as implying that all the genes with which a fertilized egg is endowed remain intact, though not necessarily expressed, throughout the life of the organism. The validity of this deduction can be assessed by considering for a moment the view that differentiation proceeds by loss or other alteration of genes. Several mechanisms are known by which such mutation can be achieved. Indeed it has been suggested that this may be the basic mechanism underlying carcinogenesis. What is more, mutations can be reversed, since microbial mutants have been shown to revert to their parent type in a manner almost reminiscent of the ability of differentiated plants and animal cells to revert to their embryonic state. However, there is one important difference. Reversion in microbes occurs normally at a rate of 1 in 10^8 cells, and even though it can greatly be increased by the use of powerful mutagenic chemicals, it remains essentially a random process. In the examples just cited, however, individual plant cells or frog cell nuclei revert not only with high frequency but, most important, return to the original state with every gene apparently correct. No mechanism of chemical mutagenesis is known which can account for such specificity, and it may be concluded that totipotency of cells does imply the intactness of genetic material during differentiation.

On the other hand it is possible that the differential amplification of certain genes, as demonstrated in the case of ribosomal RNA (Section 8-1(b)) and suggested for the haemoglobin genes in erythropoietic tissue, may play a fundamental part in development. What is important is that at least one copy of each gene is replicated unchanged at every cell division. The development of antibody-forming cells (Section 4-3(b)) appears to be an exception to this generalization. In that case, somatic mutation of certain genes may be operative.

7-2 Site of Genetic Stimulation

Having established that the cell in an organism contains all the genes which that individual is capable of expressing, the next question is clearly the following. What prevents the expression of eye genes in liver or of muscle genes in blood, and so forth? or in biochemical terms, how is myosin synthesis expressed in muscle cells but prevented in all other cells? Before considering that question in detail (Chapter 8) another point of fact can be established. It concerns the site of the stimulus that switches the genes on and off. Two cases in which a rather dramatic activation of genes occurs will be considered.

(a) Activation of Eggs

It appears from Gurdon's nuclear transplantation experiments that although the genes are in the nucleus of cells, it is the cytoplasm that controls their expression. This is most clearly shown in the following experiments (Figure 7-4). When nuclei from adult red cells (which never synthesize DNA) or brain (which rarely synthesize DNA) are transplanted into enucleated eggs, DNA synthesis, as measured by incorporation of radioactive thymidine and detected by autoradiography (Figure 3-3), begins within one

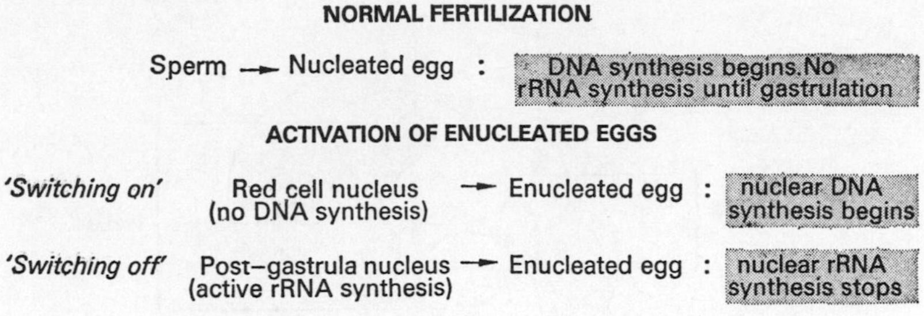

NORMAL FERTILIZATION

Sperm → Nucleated egg : | DNA synthesis begins. No rRNA synthesis until gastrulation |

ACTIVATION OF ENUCLEATED EGGS

'Switching on'   Red cell nucleus   → Enucleated egg : | nuclear DNA synthesis begins |
(no DNA synthesis)

'Switching off'   Post–gastrula nucleus → Enucleated egg : | nuclear rRNA synthesis stops |
(active rRNA synthesis)

Figure 7-4 Cytoplasmic control of genetic expression during activation of enucleated frog eggs

hour in over 70 per cent of egg cells. Moreover it is localized over the injected nuclei.

Conversely, other genes can be shown to be turned *off* by egg cytoplasm. Thus frog embryos beyond the gastrula stage synthesize much ribosomal RNA; prior to this stage they do not (Section 3-3(a)). Transplantation of post-gastrula nuclei into unfertilized eggs immediately stops their synthesis of rRNA; when gastrulation is reached once more normal synthesis of rRNA resumes.

(b) Hybrid Cells

Harris has used the technique of cellular fusion to obtain essentially similar results. In order to fuse cells, their membranes are modified by incubation in the presence of a para-influenza (Sendai) virus which has been inactivated by ultraviolet light. The cells can be from different individuals and even different species such as mouse and man. The resulting mixture contains several hybrid or heterokaryon cells, which can be distinguished by the fact that they contain more than one nucleus (Figure 7-5). The origin of the nuclei can be recognized by virtue of their size and also autoradiographically by fusing a cell line previously treated with radioactive thymidine (which labels only the nucleus) with an unlabelled one. Figure 7-5 shows a hybrid cell containing three human nuclei and two rat nuclei obtained by fusing HeLa cells, a human cancer maintained in culture for many generations, with

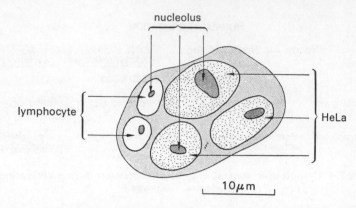

Figure 7-5 Cell fusion. Hybrid cell resulting from the fusion of rat lymphocytes with HeLa cells (human cancer cells). In this instance, three HeLa
cells have fused with two lymphocytes. The HeLa nuclei can readily be
distinguished by their size and by the fact that they have been prelabelled
with radioactive thymidine (see Figure 3-3), shown by the black dots. After
Harris (1968)

rat lymphocytes, one of the white blood cells concerned with antibody production. Harris has also fused HeLa cells (like most cancer
cells easy to maintain in culture) with rabbit macrophage (another
type of white cell), mouse cancer cells, hen and frog red cells, and
so forth.

Hybrid cells often undergo mitosis, and many can be propagated
indefinitely (Figure 6-7). Before a hybrid cell divides, all its
constituent nuclei synthesize DNA and RNA (detected autoradiographically after exposure to the appropriate radioactive
precursors or by microspectrophotometric analysis of total nucleic
acids). Thus fusion of HeLa cells (which divide rapidly and have
high rates of RNA and DNA synthesis) with hen red cells (which
do not divide and synthesize no DNA or RNA) leads to hybrid
cells showing DNA and RNA synthesis over hen nuclei as well as
over HeLa ones; this suggests that the cytoplasm, containing
some, if not exclusively HeLa material, controls the biochemical
expression of the nuclei within it (Table 7-1).

It may be argued that the effect of the cytoplasm in activating
dormant nuclei is not specific to HeLa cells, but is a reflexion of
some general 'foreignness' of the situation. However, if rabbit

Table 7-1 Nucleic acid synthesis in hybrid cells. Cells were exposed to radioactive uridine or thymidine and incorporation into RNA and DNA respectively detected autoradiographically (see Figure 3-3)[a]

| | Nuclear synthesis of | |
	RNA	DNA
Parent cell		
HeLa	+	+
Rabbit macrophage	+	−
Rat lymphocyte	+	−
Hen red cell	−	−
Hybrid cell		
HeLa–HeLa	+ +	+ +
HeLa–Rabbit macrophage	+ +	+ +
HeLa–Rat lymphocyte	+ +	+ +
HeLa–Hen red cell	+ +	+ +
Rabbit macrophage–rabbit macrophage	+ +	− −
Rabbit macrophage–rat lymphocyte	+ +	− −
Rabbit macrophage–hen red cell	+ +	− −

+ + denotes synthesis in nuclei from both parents; − − denotes synthesis in neither nucleus.

[a] From Harris (1968).

macrophage (which makes RNA but no DNA) is fused with hen red cells, only RNA synthesis, in hen and rabbit nuclei, is observed; no DNA is made (Table 7-1). In other words, if either 'parent' makes RNA or DNA, then both types of nucleus in the hybrid cells will make RNA and DNA, but if neither parent makes DNA then neither nucleus will do so either. Although the cytoplasmic signals reflect the nature of each parent, they are unspecific with regard to the type or species of nucleus which they stimulate.

7-3 Concluding Remarks

The experiments described in this chapter have established two facts: first, the genes of the embryonic nucleus remain intact during differentiation; second, their expression is determined by the nature of the surrounding cytoplasm. Hence it appears that one must look to cytoplasmic factors for an understanding of the mechanism by which protein synthesis is controlled during development.

7-4 Summary

(a) *Totipotency of Differentiated Cells*

The view that synthesis of cell-type specific proteins is due to *mutations* of *DNA* in differentiated tissue must be *rejected* on the basis of experiments which show that fully developed tissues have *totipotency*. Thus *carrot root phloem* and *tobacco pith callus* can be cultured so as to produce *complete plants*; nuclei of *adult frog intestine* when injected into *enucleated eggs* (which are incapable of further development) give rise to *normal tadpoles* and *frogs*. Moreover differentiated *bacterial spores* and *slime mould spores* germinate to reproduce fully competent vegetative cells.

(b) *Site of Genetic Stimulation*

The *stimulus* for the expression of nuclear genes is *specific* to the *cytoplasm* in which they act. This is indicated by experiments with *frog nuclei* transplanted at different stages of development into *enucleated eggs* and with *differentiated cells* of distinct species fused together to make multinucleate *hybrid cells*.

Selected Bibliography

General Reading on Mechanism of Differentiation

Bonner, J. (1965). *The Molecular Biology of Development*. Oxford University Press, Oxford

Bullough, W. S. (1967). *The Evolution of Differentiation*. Academic Press, London–New York

Clever, U. (1968). Regulation of Chromosome Function. *Ann. Rev. Genetics*, **2**, 11

Davidson, E. H. (1969). *Gene Action in Early Development*. Academic Press, London–New York

Gross, P. R. (1968). Biochemistry of Differentiation. *Ann. Rev. Biochem.*, **37**, 631

Gurdon, J. B. (1968a). Nucleic Acid Synthesis in Embryos and its Bearing on Cell Differentiation. In *Essays in Biochemistry* (Ed. P. N. Campbell and G. D. Greville), Academic Press, London–New York, Vol. 4, p. 25

Harris, H. (1968). *Nucleus and Cytoplasm*. Oxford University Press, Oxford

Paul, J. (1968). Molecular Aspects of Cytodifferentiation. *Adv. Compar. Physiol. & Biochem.*, **3**, 115

Symposium on Molecular Aspects of Differentiation (1968). *J. Cell. Physiol.*, **72**, Suppl. 1

Totipotency

Heslop-Harrison, J. (1967). Differentiation. *Ann. Rev. Plant Physiol.*, **18**, 325

Gurdon, J. B. (1966). The Cytoplasmic Control of Gene Activity. *Endeavour*, **25**, 95

Gurdon, J. B. (1967). Nuclear Transplantation and Cell Differentiation. In *Cell Differentiation, CIBA Foundation Symposium* (Ed. A. V. S. de Reuck and J. Knight), J. & A. Churchill, London, p. 65

Gurdon, J. B. (1968b). Transplanted Nuclei and Cell Differentiation. *Scientific American*, **219**, No. 6, p. 24

King, T. J. (1966). Nuclear Transplantation in Amphibia. In *Methods in Cell Physiology* (Ed. D. M. Prescott), Academic Press, London–New York, Vol. 2, p. 1

McCarthy, B. J. and B. H. Hoyer (1964). Identity of DNA and Diversity of mRNA Molecules in Normal Mouse Tissues. *Proc. Nat. Acad. Sci.*, **52**, 915

Spemann, H. (1938). *Embryonic Development and Induction*. Yale University Press, New Haven, Conn.

Steward, F. C. (1964). Growth and Development of Cultured Plant Cells, *Science*, **143**, 20

Steward, F. C., A. E. Kent and M. O. Mapes (1966). The Culture of Free Plant Cells and its Significance for Embryology and Morphogenesis. *Current Topics in Dev. Biol.*, **1**, 113

Vasil, V. and A. C. Hildebrandt (1965). Differentiation of Tobacco Plants from Single, Isolated Cells in Microcultures. *Science*, **150**, 889

Site of Genetic Stimulation

Ephrussi, B. and M. C. Weiss (1969). Hybrid Somatic Cells. *Scientific American*, **220**, No. 4, p. 26

Gurdon, J. B. and M. R. Woodland (1968). The Cytoplasmic Control of Nuclear Activity in Animal Development. *Biol. Rev.*, **43**, 233

Harris, H. (1970). *Cell Fusion*. Oxford University Press, Oxford

CHAPTER 8

Mechanism: hypotheses

It was seen in the last chapter that genetic mutation cannot be the basic mechanism of differentiation. Furthermore in nuclear transplantation and in cell fusion experiments, in which nuclear activity is turned on and off, the stimulus proved to be cytoplasmic. This stimulus was discussed somewhat as though it were the initiator of differentiation (Chapter 1). But is this justified? Nuclear transplantation and cell fusion are rather unphysiological events and it is by no means obvious that the mechanism of normal differentiation involves genetic stimulation at all. Even during fertilization of sea urchin eggs, a wave of cytoplasmic protein synthesis *precedes* the duplication of DNA (Section 3-2(a)). What was established in previous chapters is that protein synthesis, specific with regard to type and rate, is the biochemical basis underlying differentiation.

8-1 Control of Protein Synthesis

(a) *The Problem*

Whether a particular protein is made, and if so, at what rate, may be controlled not only at the genetic, transcriptional level but also at the stage of translation (Figure 8-1). In each case the controlling elements are cytoplasmic. Transcriptional control implies that the amount and kind of protein that is synthesized is proportional to the amount and type of mRNA transcribed. In other words gene activation is selective. Translational control, on the other hand, implies that individual templates of mRNA are copied into protein only as required, and that the timing of transcription is not crucial. A third possibility is that both mechanisms operate. In the case of ribosomal RNA synthesis, and possibly of mRNA also, the initial gene product appears to be larger than the finished RNA. Such 'processing' of RNA is another potential site for metabolic control.

In this chapter an attempt will be made to see which form of

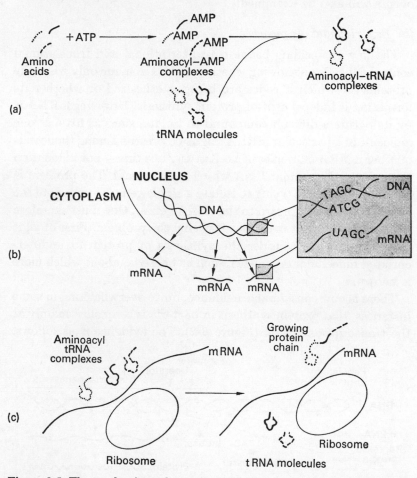

Figure 8-1 The mechanism of protein synthesis. *Amino acid activation* (a): Amino acids react with ATP in the cytoplasm of the cell to form aminoacyl–AMP complexes; the aminoacyl group is transferred to specific transfer RNA (tRNA) molecules. *Transcription* (b): Specific regions of DNA—the genes—are copied by synthesis of 'complementary' molecules of RNA. These messengers (mRNA) somehow reach the cytoplasm. *Translation* (c): Aminoacyl–tRNA complexes bind to complementary regions on mRNA, which is 'anchored' to ribosomes. Aminoacyl residues are joined through peptide bonds to form a growing protein chain

6+

control best fits the facts in differentiating systems. The relation between the factors that control protein synthesis and the signals that have been identified as stimulating certain developmental events will also be examined.

(b) Experimental Approaches

The key to deciding between transcriptional and translational control depends on showing whether mRNA is made only when the proteins for which it codes are being synthesized or whether its formation is independent of protein synthesis. However mRNA is by its nature a difficult compound to isolate, since as little as one molecule of a particular mRNA might be present among thousands of other mMRA molecules in a cell at any one time—not to mention the bulk of the cellular RNA which is ribosomal. The problem is about as difficult as trying to isolate a single gene (though note the initial success with regard to the rRNA genes). One must therefore fall back on indirect methods of solving the problem. First of all it may be useful to consider the synthesis of protein in undifferentiated unicellular organisms, such as bacteria, about which more is known.

There is now considerable evidence, quite overwhelming in some instances, that protein synthesis in bacteria is controlled mainly at the transcriptional level (Figure 8-2). The evidence is as follows.

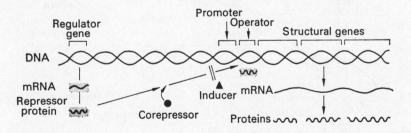

Figure 8-2 Control of transcription in microbes. The transcription of each group of structural genes is initiated by the binding of the enzyme RNA polymerase to a 'promoter' region. The system is controlled by a 're-pressor' protein which prevents transcription by binding to an 'operator' site. The association of repressor with the operator site is stimulated by low molecular weight 'co-repressors' and inhibited by low molecular weight 'inducers'. This mechanism is known as 'negative control'; in some instances the reverse mechanism ('positive control') operates

First of all, the RNA molecules corresponding to specific enzymes and groups of enzymes have been demonstrated to be made just when the relevant enzymes are being synthesized. However, no mRNA has yet been isolated in pure form. Now the synthesis of mRNA is postulated to be controlled by a specific endogenous 'repressor' which binds the corresponding region of DNA; the repressor, which appears to be a protein, allows or prevents transcription depending on whether it has attached, at a second site, an exogenous, small molecular weight inducer or repressor. Inducers are generally substrates, and repressors metabolic end products, of the enzyme under consideration. In two instances specific repressor protein has been isolated. This provides the strongest evidence yet available for the transcriptional control of protein synthesis in bacteria. In some cases translational control may be superimposed on this basic mechanism. The question to be decided now is whether protein synthesis in differentiated cells is controlled by the same process.

Four types of experimental approach have been employed. They may be summarized as (1) use of enucleated cells; (2) use of inhibitors; (3) attempts to detect specific RNA synthesis and (4) attempts to demonstrate specific masking of genes. Each technique will be described in turn, illustrating it wherever possible with the systems encountered in previous chapters.

(i) *Enucleated Cells*. If protein synthesis is shown to be controlled in the absence of the nucleus, this is clearly strong evidence in favour of translational regulation, since it implies the existence of a stable mRNA, the expression of which is under cytoplasmic control. Several experiments do just that.

The best evidence comes from some unicellular organisms called *Acetabularia*. These are giant algae which can grow to 3–5 cm in length when mature. They are morphologically and biochemically differentiated in that they possess a distinct cap, containing polysaccharides not found elsewhere in the cell, at one end of a long stalk. At the other end are small protuberances or rhizoids, one of which contains the nucleus (Figure 8-2). The nucleus is easily removed without damage to the rest of the cell simply by cutting

the relevant rhizoids off. Such cells eventually develop normal
caps. The nucleus can be transplanted back into enucleated cells.
Different species of *Acetabularia* have morphologically distinctive
caps and can be used as genetic markers in nuclear transplantation
experiments, in the way described for pigmented tadpoles in
Section 7-1(c). Thus if a nucleus of *A. mediterranea*, which has a
large umbrella-like cap, is injected prior to cap formation into
a young enucleated host such as *A. crenulata*, destined to form a

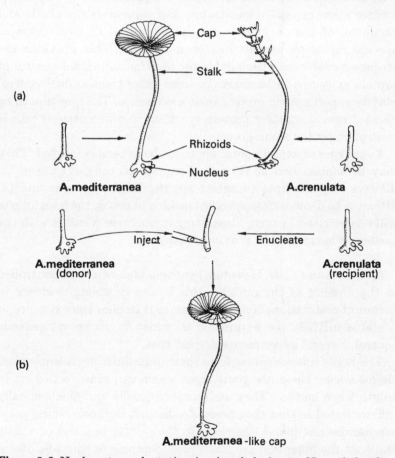

Figure 8-3 Nuclear transplantation in *Acetabularia*. (a) Normal development; (b) nucleus from *A. mediterranea* is injected into a young enucleated
A. crenulata; the resulting 'hybrid' forms an *A. mediterranea*-like cap

smaller cap, a large *A. mediterranea*-like cap is eventually formed (Figure 8-3). In other words cap morphology is determined by the nucleus. On the other hand the timing of cap formation is under cytoplasmic control. Thus if immature cells are exposed to altered environmental conditions such as the intensity of illumination prior to cap development the subsequent formation of cap can be accelerated or retarded by several weeks. This result is reminiscent of the effect of cytoplasm in the nuclear transplantation experiments discussed in Section 7-2.

Since cap formation involves the synthesis of specific proteins (enzymes involved in the synthesis of the cap polysaccharides are an example), the experiments with enucleated *Acetabularia* cells demonstrate the operation of translational control. It may be argued that because the cytoplasm contains DNA (in the chloroplasts and mitochondria), transcriptional control cannot be ruled out. However, cap morphology is specified by the nucleus, as seen above, and so are the enzymes concerned with polysaccharide synthesis in the cap. Whatever the significance of cytoplasmic DNA, it is not with cap formation.

Experiments with enucleated sea urchin eggs lead to essentially similar results. In this case, enucleation is achieved by centrifuging eggs, which separate into two halves; a heavy, yolk-containing part and a lighter half which contains the nucleus (Figure 8-4). If the enucleated part is 'activated' by pricking it with a needle, it begins to cleave and continues development right up to the blastula stage. These processes, as in the normal development of a nucleated egg, include a wave of protein synthesis. Enucleated eggs of some other species behave in a similar manner.

Although this experiment clearly demonstrates the existence of long-lived mRNA, it does not provide information regarding the regulation of its translation into protein. What is known is that between the time the RNA is made (in the developing oocyte) and its first expression immediately after fertilization or activation, the machinery for protein synthesis is present. Hence the RNA must be 'masked' in some way. It occurs in the cell cytoplasm, perhaps associated with protein in ribosomal-like particles, termed 'informosomes'. How the protein is involved in masking is as yet

6*

not known. An example in higher organisms is provided by the 'naturally' enucleated mammalian reticulocyte (Section 4-3(*a*)). The synthesis of haemoglobin, which begins in the erythroblast, is not only maintained in the reticulocyte but is susceptible to regulation by the amount of haem (the prosthetic group of haemoglobin) present.

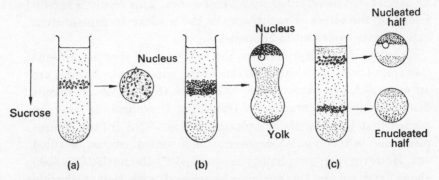

Figure 8-4 Centrifugation of sea urchin eggs to give nucleated and enucleated halves. Eggs are introduced half-way down a centrifuge tube containing an increasing concentration of sucrose (a). During centrifugation, eggs 'stretch' into nucleated and yolk-containing halves (b) which eventually separate and either float to the top or sink to the bottom of the tube (c). After Tyler (1967)

One cannot but conclude that in the situations just cited translational control is operative. The nature of the regulating mechanism is not clear; in the cases where translational control has been demonstrated in microorganisms, rRNA, tRNA and other molecules may be involved.

(ii) *Actinomycin D*. Throughout this book a distinction has been drawn between activation and synthesis of enzymes, by referring to the use of inhibitors of protein synthesis. Two widely-used compounds are puromycin and cycloheximide (actidione) (Figure 8-5); they are specific in that DNA and RNA synthesis, and most metabolic enzymes, are unaffected by the presence of the inhibitor at low concentration. In other words the inhibition appears to be at the translational stage. An indirect inhibitor of protein synthesis that has been much used is actinomycin D (Figure 8-5). In this case RNA synthesis is prevented and there is quite good evidence to show that it is the transcription of DNA that is inhibited. The cessation of protein synthesis is therefore a secondary effect. The reason protein synthesis stops is because mRNA is generally not a stable molecule but is subject to continual breakdown and re-

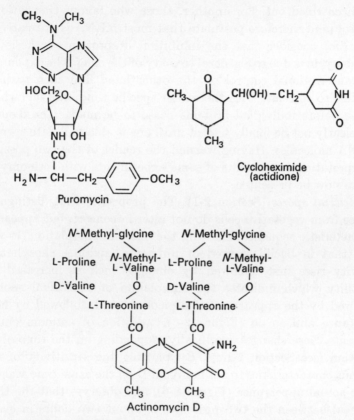

Figure 8-5 Some inhibitors of protein synthesis

synthesis ('turnover'). If synthesis is blocked by actinomycin D, protein synthesis will continue only as long as sufficient mRNA molecules remain. When actinomycin D is added to whole animals or to cells in culture, individual proteins cease to be synthesized at different times. Those which are affected first are said to be coded by mRNA that is 'short-lived' while those which are affected later are said to have a correspondingly 'longer-lived' mRNA. In fact actinomycin D has been used to estimate the longevity of specific mRNA molecules. However this interpretation is not universally accepted. For one thing, it has not been proved that it is destruction of mRNA that limits its activity; 'masking' or inhibition has not been ruled out. For another, those who favour translational control (and therefore postulate that most mRNA is of relatively long life) consider that the inhibition of protein synthesis by actinomycin is due to a general toxicity of the drug. The exponents of transcriptional control on the other hand find such toxicity difficult to correlate with the rather specific time intervals which elapse before individual proteins cease to be made. The dispute can clearly not be finally settled until one is able to isolate specific mRNA molecules. Having warned the reader of the two possible interpretations, the results of some experiments with actinomycin D can now be presented.

Bacterial spores (Section 2-1). The properties that distinguish spores from vegetative cells do not alter in concert, but appear in characteristic sequence following the onset of sporulation. It was seen that in bacilli starved of nutrients, alkaline phosphatase activity rises first, followed by changes such as increased re-fractility which indicates the completion of individual spores, followed by the appearance of dipicolinic acid, followed by heat resistance and so on (Figure 2-5). Addition of actinomycin D prevents these changes specifically, depending on the time of its addition (see Section 2-1(c)). By plotting the sensitivity of the various characteristics to actinomycin D on the same time scale as their actual appearance (Figure 8-6), one observes that the time interval between the two processes is about two hours in every case. If the onset of insensitivity, or commitment, is taken to imply the beginning of transcription, one may conclude that the mRNA molecules which code for each event are formed sequentially and have a lifetime of at least two hours.

This is rather long compared with the estimated average lifetime of mRNA in vegetative cells, which is generally only a few minutes. Either the interpretation of actinomycin sensitivity in

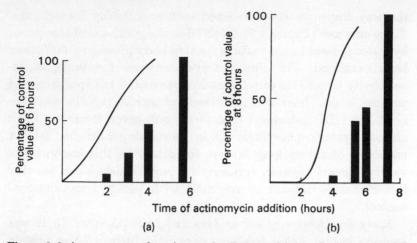

Figure 8-6 Appearance of actinomycin D-insensitivity during bacterial sporulation. The continuous curve represents the amount of alkaline phosphatase (a) or refractility (b) that is eventually achieved when actinomycin D is added at the time intervals shown. The shaded bars show the appearance of alkaline phosphatase or refractility in control cultures without actinomycin D. Note that in each case actinomycin D-insensitivity precedes expression by about two hours. From Sterlini and Mandelstam (1969)

terms of mRNA is wrong or there is a marked difference between the mRNA of vegetative and spore cells. Perhaps the mRNA of developing spores is masked in the way that mRNA of sea urchin eggs is (see above); in other words, regulation of translation is indicated. The sequential activation of specific genes, on the other hand, implies transcriptional control. DNA–RNA hybridization experiments carried out on vegetative and sporulating cells confirm that discrete parts of the genome are expressed at different times. The mechanism by which the sequence of gene activity is specified is not clear. It may be similar to that of sequential induction of enzymes in vegetative cells, in which the product of one enzyme is the inducer for the next and so on. Sensitivity to inhibitors of protein synthesis continues right up to the time of expression, which implies that translational events, at least, are not very much slower in spores than in vegetative cells.

Slime moulds (Section 2-2). The appearance of specific enzymes in *D. discoideum* is sensitive to inhibition by actinomycin D in just

the way the processes associated with sporulating bacteria are.
Three enzymes (Figure 2-7) involved in the synthesis of the carbohydrates present in the mature fruiting body (Section 2-2(b)) have
been examined. The time interval between the onset of insensitivity to actinomycin D (transcription) and the appearance of
enzyme is one hour for trehalose-6-phosphate synthetase, five
hours for UDP galactosyl transferase and seven hours for UDP
glucose pyrophosphorylase. As in sporulating bacteria, mRNA
molecules of rather long life are indicated. In this instance the
varying period between transcription and translation is another
factor that is difficult to reconcile with purely transcriptional
control.

Early development of sea urchins and frogs (Chapter 3). It was
seen in Chapter 3 that fertilized eggs make little RNA or protein
until late cleavage and it is therefore not surprising that eggs can
develop considerably in the presence of actinomycin D. Nevertheless the wave of protein synthesis induced by fertilization of sea
urchin eggs occurs in the presence of actinomycin D, and thus is
indicative of stable, 'masked' mRNA. Moreover, as with enucleated sea urchin eggs, the potentiality for protein synthesis is
present up to a relatively advanced stage of early development.
Thereafter the drug becomes increasingly toxic. For example it
prevents the formation of the 'hatching enzyme' in sea urchin
gastrulae and the synthesis of the tail-resorbing and urea cycle
enzymes in metamorphosing tadpoles.

Development of specific functions in other systems (Chapters 5
and 6). Generally speaking, whenever the appearance of enzymic
activity accompanying a developmental event is prevented by an
inhibitor of protein synthesis, it is also susceptible to inhibition by
actinomycin D. The systems do not lend themselves quite so well
to study of the commitment, or potentiality period, as do bacterial
spores and slime moulds, but some similar experiments have been
carried out. In the case of lactating mammary tissue in culture
(Section 5-1(b)), for example, actinomycin D added at the same
time as insulin (which is one of the hormones required for development) abolishes the synthesis of specific enzymes, whereas if
added some hours later the drug has no effect; this has been taken

to indicate the presence of a long-lived mRNA. The synthesis of haemoglobin induced by erythropoietin in culture (Section 5-1(b)) is inhibited by actinomycin D. As with lactating mammary tissue, synthesis of DNA and of a small amount of protein appears to precede actinomycin-sensitive synthesis of RNA. Addition of actinomycin D to insect larvae during pupation abolishes 'puffing' and the associated RNA synthesis in salivary gland chromosomes (Section 5-1(d)). Since protein synthesis continues for some time, participation of long-lived mRNA has been inferred.

A good example of the use of actinomycin D in determining the longevity of individual mRNA molecules which argues against its action as a general cell poison, is its effect on the induction of various catabolic enzymes in the liver of rats fed a diet high in hydrolysed casein (see Section 6-2(a) and Figure 6-5). By measuring the period from the beginning to the end of insensitivity, the lifetime of 3 mRNA molecules has been deduced. That for tyrosine transaminase was found to be less than three hours, for serine dehydratase 6–8 hours, and for ornithine transaminase 18–24 hours. Note that the mRNA corresponding to tryptophan pyrrolase has been estimated to have a life of more than two weeks, while that for rRNA has been put at several days.

Although it is true that the appearance of these enzymes is a typically adaptive phenomenon, there is little reason to believe that the basic mechanism of protein synthesis is any different in differentiation (see Section 4-2(b)). As mentioned above, the mechanism by which translation of mRNA molecules is specifically terminated is not clear. Certainly cells contain RNAse but that such degradative enzymes have specificity both for recognition of a particular mRNA and for the speed of its degradation seems unlikely. On the other hand specific degradation of proteins, such as that of tadpole tail collagen (Section 3-3(b)) does account for certain developmental events.

The effects of actinomycin D are, in conclusion, indicative of both forms of control. The specificity in the onset of actinomycin D-insensitivity (that is, mRNA synthesis) suggests transcriptional control. The rather long and varying intervals before the mRNA is expressed, on the other hand, point to translational regulation.

(iii) *Specific RNA Synthesis*. It has already been pointed out that the most direct way of investigating the control of protein synthesis is to see whether mRNA specific to any one protein is made at a particular time or not; in the case of multicellular organisms this has not yet proved possible; an exception is the synthesis of haemoglobin mRNA in reticulocytes. An experiment purporting to show the synthesis of mRNA specific for DOPA decarboxylase (Section 5-1(b)) in insect larvae has not been generally accepted. Instead some indirect evidence has been obtained which points to the transcriptional control of protein synthesis. Two situations have been investigated, each of which reveals a regional as well as a temporal control of RNA synthesis.

The first concerns the 'puffing' of polytene chromosomes in insect larvae and its associated RNA synthesis (Figure 5-6), which was discussed in Chapter 5. The fact that the regions of the chromosomes which puff vary somewhat from tissue to tissue, coupled with the fact that puffs appear at distinct times after ecdysone administration, is suggestive of a certain specificity of RNA synthesis. However it is most unlikely that all the enzymes required for the functioning of a specific tissue are clustered in regions, each large enough to detect visually and hence covering many hundreds of genes, and it is therefore difficult to correlate puffing and the associated RNA synthesis with specific activation of individual genes. Moreover the specific proteases of the salivary secretion continue to be made when puffing in these cells is stopped by actinomycin D.

The second approach concerns the control of RNA synthesis during embryonic development. As was mentioned in Section 3-3(a), dRNA, tRNA and rRNA begin to be made at different times after fertilization (Figure 3-5). The same is true of RNA synthesis during the germination of seeds (Section 2-4(b)). Hence gene expression is selective with regard to the class of RNA that is made; however this does not imply that it is necessarily true of individual mRNA molecules. There is regional specialization in that rRNA is made in the nucleolus (Section 3-3(a)); rRNA synthesis continues to be nucleolar throughout adult life.

An interesting situation with regard to rRNA synthesis has

been elegantly exploited by D. D. Brown (Figure 8-7). During the intensive rRNA synthesis that occurs in oocytes the nucleolus is replicated until some 1000 nucleoli are present in the nucleus (called the germinal vesicle at this stage). By isolating nucleoli it is possible to analyse the DNA they contain. It turns out to be different in buoyant density and hence in base composition from the main non-nucleolar DNA and it codes predominantly for rRNA. In this instance, then, a specific group of genes, sometimes referred to as the nucleolar organizer, which correspond to the

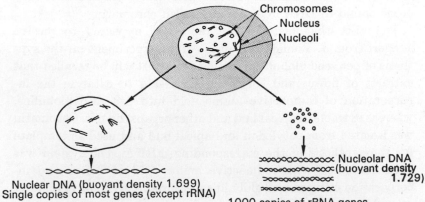

Figure 8-7 Isolation of rRNA genes. Nucleoli are dissected out of the nucleus ('germinal vesicle') of an oocyte, leaving the bulk of the nuclear (chromosomal) DNA behind. Nucleolar DNA, which has a buoyant density different to that of bulk nuclear DNA, can be shown by hybridization experiments (Figure 8-9) to code for rRNA. Drawn to illustrate the results of Brown and Dawid (1968)

components of rRNA, has been isolated. The synthesis of nucleolar DNA occurs without replication of the rest of the genome. Such amplification of specific genes may be involved in haemoglobin synthesis during erythropoiesis (Section 4-3(a)) and may prove to be an important mechanism in other developmental processes also.

The fact that many 'developmental' hormones like thyroxine and ecdysone stimulate nuclear RNA synthesis (see Section 5-1(e) and Figure 5-7) at the same time as they initiate protein synthesis, may be taken as evidence in favour of transcriptional control during differentiation.

7

(iv) *Masking of Genes*. If protein synthesis in multicellular organisms is controlled at the stage of transcription, there must be some mechanism by which certain genes are 'activated' and the remainder kept 'repressed' or 'masked' (Figure 8-9). In the case of bacteria the isolation of a repressor protein that functions in this way has already been noted. Such specific molecules have not yet been identified in multicellular organisms. Instead investigators have looked at the types of compound present in the nucleus which might function as gene repressors. The basic nuclear proteins known as histones have been most widely studied since they occur bound to DNA in a complex called chromatin.

The view that histones are repressors appeared to receive support from J. Bonner as a result of an experiment on the synthesis of pea seed globulins (Section 2-4(*b*)). It will be recalled that extracts of flower and cotyledon were said to catalyse the incorporation of radioactive amino acid into pea seed globulins, whereas extracts of apical bud and other organs did not. Chromatin was isolated from cotyledon and apical bud and used as template for the synthesis of the corresponding mRNA. The system was then incubated with radioactive amino acid as before and incorporation into pea seed globulins measured. Cotyledon chromatin caused formation of pea seed globulins but apical bud chromatin did not (Table 8-1). However when pure DNA was isolated from the chromatin by removal of histones and the experiment repeated, cotyledon and apical bud DNA were found to be equally effective. Bonner concluded that histones are the specific repressors of DNA transcription. Unfortunately this experiment has not been repeated and it would be prudent for the time being to regard the result as no more than a model of the properties expected of a repressor molecule. But that histones bind DNA in such a way as to render it generally less effective as a template for RNA synthesis has been fully confirmed.

The idea that histones are involved in gene repression has been taken up by other workers who have shown that it may be chemical modification of histones, rather than their total removal, that activates genes. Four types of change have been investigated; acetylation, methylation, phosphorylation and reduction (Figure 8-8).

Table 8-1 Bonner's pea seed globulin experiment. Chromatin, or DNA isolated therefrom, was used as a template for protein synthesis *in vitro*. Incorporation of a radioactive amino acid into pea seed globulins, separated from total protein by an immunological technique (Section 4-3(*b*)), was measured[a]

Template for RNA synthesis	Incorporation into pea seed globulins (as percentage incorporation into total protein)
Cotyledon chromatin	7.1
Apical bud chromatin	0·12
Cotyledon DNA	0·46
Apical bud DNA	0·45

[a] This experiment, taken from Bonner (1965), *The Molecular Biology of Development*, Oxford University Press, Oxford, pp. 9 and 23, has been criticized on technical grounds and the reader is advised to consider the result as a model, rather than as an observed fact, of the properties expected of repressor molecules.

If lymphocytes are incubated with radioactive acetate and the chromatin examined by autoradiography of the nucleus, the 'diffuse' or active (in terms of RNA synthesis) regions of chromatin take up label, whereas the 'dense' or inactive areas do not. Similarly, during liver regeneration following removal of part of the organ, acetylation of histones is highest when RNA synthesis is maximal. Isolated histones added to a DNA-dependent RNA synthesizing system are progressively less inhibitory, the more acetylated they are.

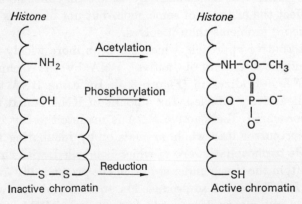

Figure 8-8 Chemical modification of histones

Several workers have isolated histones that are methylated to various degrees, but so far the physiological role of methylation is unclear.

Phosphorylation has been found to affect histones rather like acetylation. Thus histones extracted from diffuse chromatin of thymus contains more phosphate (on serine residues) than that from dense chromatin. Most of the phosphate is bound to a particular class of histone rich in lysine, called F1. In regenerating liver the phosphate content of F1 histone reflects the rate of DNA synthesis, the changes in phosphate content preceding that of DNA by up to one hour. Chemical phosphorylation of F1 histones renders them less able to inhibit DNA polymerase, while phosphorylation of F3 histones (an arginine-rich fraction) has the same effect on RNA polymerase.

The –SH:S–S (cysteine : cystine (Figure 2-2)) ratio of F3 histones appears to fluctuate in concert with morphological changes in sea urchin eggs following fertilization. Moreover fully reduced F3 histone of thymus is a less effective inhibitor of DNA-primed RNA polymerase than is the oxidized form.

Although none of these modifications has yet been correlated with the synthesis of specific proteins, it does suggest that histones in one form or another may play a rather general role in gene activation. What is doubtful is whether histones are in fact heterogeneous enough to account for the specific masking of individual genes. The mechanism by which enzymic modification might affect the histones of some genes but not of others appears to pose more problems than it solves.

An alternative approach, which has been more widely applied, is to study the extent of 'active' DNA by the technique of DNA:RNA hybridization (Figure 8-9). By using RNA that has been radioactively labelled, the amount of RNA bound to DNA can be measured. Radioactive RNA is made either by injecting labelled precursor into whole animals or by incubating the four nucleoside triphosphates, one of which is labelled, with chromatin isolated from the tissue under study, in the presence of RNA polymerase. Labelled RNA so produced is purified and then added in increasing amounts to denatured, single stranded DNA, which has been extracted from the relevant tissue, until a 'saturation' value is obtained. For most tissues this is approximately 5 to 10 per cent of its DNA. If purified DNA is used as a template instead of chromatin for the preparation of RNA by the incubation technique, the saturation value is much higher. One may conclude that, subject to certain provisos, the saturation value corresponds to the amount of DNA in the chromatin that is actively being transcribed.

If the hybridization is carried out in the presence of non-radioactive RNA, the amount of labelled RNA that is bound to DNA is reduced. Non-radioactive RNA prepared from the same tissue as the labelled RNA is more inhibitory than RNA prepared from another tissue, showing that distinct regions of the genome

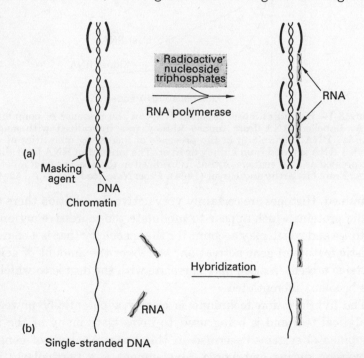

Figure 8-9 Hybridization of RNA transcribed by chromatin. (a) Formation of labelled RNA; only unmasked regions of chromatin are transcribed. (b) Labelled RNA binds only to those regions of DNA (extracted from chromatin and dissociated into single strands) from which it was originally copied

are transcribed in different tissues (Figures 8-9 and 8-10). But
since RNA from another tissue does interfere appreciably, there
must be considerable overlap of active genes. This is what one
might expect from the existence of common metabolic pathways
in differentiated tissues.

Using such hybridization techniques, the nature of the compounds in chromatin which mask gene expression has been

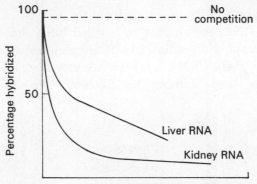

Figure 8-10 Hybridization of labelled RNA in the presence of competing
RNA. Labelled RNA from mouse kidney was hybridized with single-
stranded DNA (Figure 8-9) in the presence of increasing quantities of unlabelled RNA isolated from kidney or liver. The amount of RNA hybridized
is expressed as a percentage of that hybridized in the absence of competing
RNA. From McCarthy and Hoyer (1964), *Proc. Nat. Acad. Sci. U.S.*, **52**, 915

examined. Histones are certainly very active. In addition there are
acidic proteins which appear to modulate the repressive action of
histones and which may account for the specificity that is a characteristic feature of gene activation. Moreover the amount of acidic
proteins present in a cell may correlate with the degree to which it
has become differentiated.

The hybridization technique is clearly a potentially powerful
analytical tool and is being used to investigate many of the developmental systems described in this book. Analysis of genetic
expression during embryonic development is a particularly important situation, since regional and temporal specialization occur
simultaneously. In general the results show that regions of DNA
do appear to be rather specifically masked in different cells.

None of the experiments discussed in this section has yet revealed the presence of a repressor as specific as the bacterial proteins mentioned earlier. On the other hand some results, especially from hybridization experiments, do point rather strongly towards the selective, transcriptional control of gene expression.

8-2 Controlling Signals

(a) *Nature*

The relation between the agents that control transcription and translation, and the initiators and inducers of differentiation discussed in earlier chapters must now be considered. From what has been said, it appears likely that proteins of one sort or another are the molecules that interact directly with the genes of microbes and higher organisms. There is strong evidence that in bacteria such proteins combine with a small molecule, the inducer or repressor, at a second specific site. The induction of adaptive events in higher organisms, such as the synthesis of specific degradative enzymes by amino acids (Section 6-2(*a*)), is presumably by a similar mechanism. Whether inducers of development, such as the hormones that elicit specific enzyme synthesis (Section 5-1(*b*)), act in the same manner is not known. If they do, it presupposes that control is transcriptional in this instance and that the proteins which mask the relevant genes are of distinct structure in different tissues. Less is known about translational control, even in microbes. Ribosomes appear to play a part and one might therefore note that an alternative mode of action of hormones that induce enzyme synthesis may be correlated with the increased synthesis of rRNA that has been observed (Section 5-1(*e*)).

In some instances, cytoplasmic proteins have been shown to stimulate gene activity. The synthesis of DNA which is initiated in the nucleus of frog and sea urchin eggs by fertilization, or that which follows the formation of hybrid cells, is accompanied by the passage of proteins from cytoplasm to nucleus. Whether the proteins are enzymes, such as DNA polymerase, is not clear. Their function may well be concerned with the regulation of specific genes. Furthermore some form of nuclear activation may be a prerequisite for the expression of individual genes. For example, rupture of the nuclear membrane, such as occurs during mitosis, may be necessary for some specific cytoplasmic components to come into contact with chromosomes; activation of distinct genes could follow during interphase when chromatin becomes un-condensed. In other words, a round of cell division may have to precede each re-programming of genes at certain stages of develop-ment (see (*b*) below).

That chemical transmitters are involved in aspects of differentia-tion such as the control of organ size is clear from experiments in which one of the kidneys of an animal is removed; growth of the other is stimulated until its functional capacity is exactly twice that previously. The same is true if one of the livers of a pair of parabiotic (having a common circulation) twins is removed. This type of organ regeneration, it may be noted, contrasts sharply with the uncontrolled growth of cancers (Section 6-1).

(*b*) *Transmission*

How initiators, which may be synthesized only in certain cells of developing organs or organisms (as in the secretion of cyclic AMP by slime moulds (Section 2-2(*c*)), or at certain times (as in the production of 'developmental' hormones such as thyroxine or ecdysone (Section 5-1(*b*)), are transmitted to other cells is not known. Substances like hormones are almost by definition capable of traversing between cells, but nucleotide complexes such as that postulated as initiator of chondrogenesis (Section 4-1(*c*)) or cyclic AMP would have more difficulty in penetrating cellular mem-branes. In this respect it is interesting to note that in various epi-thelial tissues of insects, amphibia and mammals there appears to be free passage of electrolytes and other molecules including pro-teins across particular regions of the cell boundary. If chemical gradients were produced in this way, quantitative differences as well as qualitative similarities between distant cells could result. The phenomenon of contact inhibition discussed in Section 6-2(*b*) is another example of the way in which some form of intercellular communication may be achieved. That cells of different organs 'recognize' each other is shown by experiments in which chick or mouse embryos are dissociated into discrete cells; these re-aggregate in culture to form clusters of the original cell types.

A further mechanism by which the composition of cells may be altered during development is by unequal division of the cyto-plasm at mitosis. In this way the concentration of key substances such as proteins, low molecular weight metabolites or ions could be delicately varied.

8-3 Concluding Remarks

It will have become apparent to the reader that both mecha-nisms of control of protein synthesis—transcriptional and trans-lational—are likely to be involved in differentiation. Results of experiments with enucleated cells indicate translational control, those on specific RNA synthesis and gene activation imply transcriptional regulation, whereas the use of actinomycin D leads to the conclusion that both mechanisms are operative. The time is not yet right for reaching final conclusions about the physiological significance of each type of control, and all one may do is speculate.

What seems plausible is the following: gene activation *is* specific; sometimes single genes are concerned, as in the synthesis of alkaline phosphatase by sporulating bacteria; sometimes whole pathways are turned on, as in the synthesis of the urea cycle enzymes during amphibian metamorphosis. Whether expression (translation) follows closely upon transcription as in many hormonally-induced events, or whether there is a longer interval which is susceptible to external variation, as in the formation of caps in enucleated *Acetabularia* or in the synthesis of proteins following the fertilization of eggs, depends on the individual situa-tion. It is the genetic, transcriptional control that regulates the basic complement of enzymes in differentiated tissues, whereas the modulation of these activities, by hormones and by other agents, may be effected by mechanisms that are essentially translational.

The other major problem that remains concerns the maintenance of the differentiated state in the cells of a particular tissue through-out the life of an organism. The problem is essentially that of the stable inheritance of different phenotypic characteristics without an alteration in genetic content (see Section 6-1). Extrachromosomal heredity has been observed in the unicellular amoeba *Paramecium* and in certain yeasts and other microbes, and it is possible that elucidation of these phenomena will ultimately have a bearing on the problem of differentiation in higher organisms.

8-4 Summary

(a) *Control of Protein Synthesis*

Protein synthesis is susceptible to control at the *transcriptional* stage of gene activation (DNA → RNA) or at the *translational* stage of template utilization (RNA → protein). No intermediate (messenger) RNA has yet been isolated from higher organisms and it is therefore not possible to choose definitively between the two alternatives. Four experimental approaches have been used to investigate the two possibilities.

(1) Experiments with cells, such as *Acetabularia*, which have been artificially *enucleated* show that protein synthesis is susceptible to regulation in the absence of nucleus. Hence *translational* control must be operative.

(2) The use of *actinomycin D*, which prevents DNA-primed RNA synthesis, has shown that the synthesis of many proteins is sensitive to inhibition only at certain times prior to expression. Such *temporal* control of RNA synthesis is comparable with *transcriptional* regulation but the rather long and specific time intervals that elapse before expression of the mRNA takes place indicate *translational* control.

(3) In two situations specific RNA synthesis indicative of *transcriptional* control has been observed. First, the '*puffing*' of insect salivary gland chromosomes at specific *sites* and *times* is accompanied by *RNA synthesis* at the same sites and times. Second, during the development of amphibian eggs, *synthesis* of *dRNA*, *tRNA* and *rRNA* begins at *different times* after fertilization; there is *regional* control insofar as *rRNA* is made in the *nucleolus*.

(4) Selective expression of individual genes implies the presence

of substances specifically *masking* the remainder. Two types of compound have been implicated. (*a*) *Basic proteins* of the nucleus known as *histones* which inhibit DNA transcription and DNA replication; *chemical modification* of particular types of histone may provide some specificity. (*b*) *Acidic nucleoproteins* which also mask genes and may be more heterogeneous. In no case have repressors of discrete genes yet been isolated. Experiments on *hybridization* of RNA with DNA indicate that the *regions of DNA* that are *masked* vary from tissue to tissue. This again points to *transcriptional* regulation.

(b) Nature of Controlling Signals

The *nature* of the *signals* that exert transcriptional or translational control, their relation to known inducers of differentiation and their *propagation* through the cells of a developing organ remain to be elucidated.

Selected Bibliography

General Reading on Mechanism of Differentiation
See Chapter 7

General Reading on Control of Protein Synthesis

Bock, R. M. (1969). Translation as a Control Factor in Genetic Expression. In *Exploitable Molecular Mechanisms and Neoplasia*. The Williams and Wilkins Company, Baltimore, Md., p. 191

Bretscher, M. S. (1968). How Repressor Molecules Function. *Nature*, **217**, 509

Cohen, N. R. (1966). The Control of Protein Synthesis. *Biol. Rev.*, **41**, 503

Epstein, W. and J. R. Beckwith (1968). Regulation of Gene Expression. *Ann. Rev. Biochem.*, **37**, 411

Gross, P. R. (1968). The Control of Protein Synthesis in Embryonic Development and Differentiation. *Current Topics in Dev. Biol.*, **2**, 1

Harris, H. (1968). *Nucleus and Cytoplasm*. Oxford University Press, Oxford

Haynes, R. H. and P. C. Hanawalt (intr.) (1968). *The Molecular Basis of Life, Part III. Gene Action in Protein Synthesis*. Readings from *Scientific American*. W. H. Freeman & Co., San Francisco, California

Hendler, R. W. (1968). *Protein Biosynthesis and Membrane Biochemistry*. John Wiley & Sons, London–New York

McFall, E. and W. K. Maas (1968). Regulation of Enzyme Synthesis in

Micro Organisms. In *Molecular Genetics*, Part 2 (Ed. J. H. Taylor), Academic Press, London–New York

Pitot, H. C. (1968). Metabolic Regulation in Metazoan Systems. In *Molecular Genetics*, Part 2 (Ed. J. H. Taylor), Academic Press, London–New York

San Pietro, A., M. R. Lamborg and S. T. Kenney (Ed.) (1968). *Regulatory Mechanisms for Protein Synthesis in Mammalian Cells*. Academic Press, London–New York

Stent, G. S. (1964). The Operon: On Its Third Anniversary. *Science*, **144**, 816

Sueoka, N. and T. Kano-Sueoka (1970). Transfer RNA and Cell Differentiation. *Progress in Nucleic Acid Research*, **10**, 23

Warner, J. R. and R. Soeiro (1967). Direction of Protein Synthesis by RNA. In *Macromolecular Synthesis and Growth* (Ed. R. A. Malt), J. & A. Churchill, London, p. 31

Enucleated Cells and Actinomycin D

Brachet, J. (1968). Synthesis of Macromolecules and Morphogenesis in *Acetabularia*. *Current Topics in Dev. Biol.*, **3**, 1

Clever, U. (1969). Chromosome Activity and Cell Function in Polytenic Cells. *Exp. Cell Res.*, **55**, 306, 317

Gibor, A. (1966). Acetabularia: A Useful Giant Cell. *Scientific American*, **215**, No. 5, p. 118

London, I. M., A. S. Tavill, G. A. Vanderhoff, T. Hunt and A. I. Grayzel (1967). Erythroid Cell Differentiation and the Synthesis and Assembly of Hemoglobin. In *Control Mechanisms in Developmental Processes* (Ed. M. Locke) Academic Press, London–New York, p. 227

Pitot, H. C., C. Peraino, C. Lamar and A. L. Kennan (1965). Template Stability of Some Enzymes in Rat Liver and Hepatoma. *Proc. Nat. Acad. Sci.*, **54**, 845

Roth, R., J. M. Ashworth and M. Sussman (1968). Periods of Genetic Transcription required for the Synthesis of Three Enzymes during Cellular Slime Mold Development. *Proc. Nat. Acad. Sci.*, **59**, 1235

Spirin, A. S. (1966). On 'Masked' Forms of Messenger RNA in Early Embryogenesis and other Differentiating Systems. *Current Topics in Dev. Biol.*, **1**, 1

Sterlini, J. M. and J. Mandelstam (1969). Commitment to Sporulation in *Bacillus Subtilis* and its Relationship to Development of Actinomycin Resistance. *Biochem. J.*, **113**, 29

Tyler, A. (1967). Masked Messenger RNA and Cytoplasmic DNA in Relation to Protein Synthesis and Processes of Fertilisation and Determination in Embryonic Development. In *Control Mechanisms in Developmental Processes* (Ed. M. Locke), Academic Press, London–New York, p. 170

Specific RNA Synthesis

Brown, D. D. (1967). The Genes for Ribosomal RNA and their Transcription During Amphibian Development. *Current Topics in Dev. Biol.*, **2**, 48

Brown, D. D. and I. B. Dawid (1968). Specific Gene Amplification in Oocytes. *Science*, **160**, 272

Hadjiolov, A. A. (1967). Ribonucleic Acids and Information Transfer in Animal Cells. *Progress Nucleic Acid Research*, **7**, 195

Nemer, M. (1967). Transfer of Genetic Information During Embryogenesis. *Progress Nucleic Acid Research*, **7**, 243

Gene Masking

Busch, H. (1965). *Histones and other Nuclear Proteins*. Academic Press, London–New York

Hnilica, L. S. (1967). Proteins of the Cell Nucleus. *Progress Nucleic Acid Research*, **7**, 25

Hoyer, B. H. and R. E. Roberts (1968). Studies of Nucleic Acid Interactions using DNA-Agar. In *Molecular Genetics* (Ed. J. H. Taylor), Academic Press, London–New York

Ord, M. G. and L. A. Stocken (1971). The Nucleus. In *Fundamentals of Cell Biology* (Ed. E. E. Bittar), John Wiley & Sons, New York–London

Paul, J. and R. S. Gilmour (1968). Organ-Specific Masking of DNA in Differentiated Cells. *Symp. Int. Soc. Cell. Biol.*, **7**, 135

Spiegelman, S. (1964). Hybrid Nucleic Acids. *Scientific American*, **210**, No. 5, p. 48

Walker, P. M. B. (1969). The Specificity of Molecular Hybridization in Relation to Studies on Higher Organisms. *Progress Nucleic Acid Research*, **9**, 301

Pogo, B. G. T., A. O. Pogo, V. G. Allfrey and A. E. Mirsky (1968). Changing Patterns of Histone Acetylation and RNA Synthesis in Regeneration of the Liver. *Proc. Nat. Acad. Sci.*, **59**, 1337

Stellwagen, R. H. and R. D. Cole (1969). Chromosomal Proteins. *Ann. Rev. Biochem.*, **38**, 951

Controlling Signals

Beale, G. H. (1969). The Role of the Cytoplasm in Heredity. In *The Biological Basis of Medicine* (Ed. E. E. Bittar and N. Bittar), Academic Press, London–New York, Vol. 4, p. 81

Crick, F. (1970). Diffusion in Embryogenesis. *Nature*, **225**, 420.

Furshpan, E. J. and D. D. Potter (1968). Low Resistance Junctions Between Cells in Embryos and Tissue Culture. *Current Topics in Dev. Biol.*, **3**, 95

Gaze, R. M. (1967). Growth and Differentiation. *Ann. Rev. Physiol.*, **29**, 59

Grobstein, C. (1964). Cytodifferentiation and its Controls. *Science*, **143**, 543

Gurdon, J. B. and M. R. Woodland (1968). Cytoplasmic Control of Nuclear Activity in Animal Development. *Biol. Rev.*, **43**, 233

Heslop-Harrison, J. (1967). Differentiation. *Ann. Rev. Plant Physiol.*, **18**, 325

Moscona, A. A. (1960). Patterns and Mechanisms of Tissue Reconstruction from Dissociated Cells. In *Developing Cell Systems and Their Control* (Ed. D. Rudnick). The Ronald Press Co., New York, p. 45

de Reuck, A. V. S. and J. Knight (Ed.) (1967). *Cell Differentiation. CIBA Foundation Symposium.* J. & A. Churchill, London
Sager, R. (1965). Genes Outside Chromosomes. *Scientific American,* **212,** No. 1, p. 71
Szent-Gyorgyi, A. (1968) *Bioelectronics.* Academic Press, London–New York
Wessels, N. K. and W. J. Rutter (1969). Phases in Cell Differentiation. *Scientific American,* **220,** No. 3, p. 36
Wolff, E. (1968). Specific Interactions Between Tissues During Organogenesis. *Current Topics in Dev. Biol.,* **3,** 65
Wolstenholme, G. E. W. and J. Knight (Ed.) (1969). *Homeostatic Regulators. CIBA Foundation Symposium.* J. & A. Churchill, London

Questions

Chapter 1

1. What do you understand by differentiation, commitment and initiation?
2. Give some examples of differentiation in animals, plants and microbes.
3. To what extent are the processes of adaptation and differentiation distinct?
4. Discuss the statement that the function of proteins is either structural or enzymic.
5. How may proteins be considered to be the basic factors underlying differentiation?
6. Discuss what molecules other than proteins might warrant consideration as basic factors involved in differentiation.

Chapter 2

1. Discuss some of the features that distinguish spore cells from their vegetative counterparts. To what extent can they be explained in biochemical terms?
2. In what way has commitment during sporulation been investigated?
3. Describe how the life cycle of a slime mould differs from that of a bacterial spore in relation to cellular differentiation.
4. Discuss the evidence on which the idea of a chemical initiator of slime mould aggregation is based.
5. Compare the development of a plant with that of an animal.
6. What biochemical changes take place when seeds begin to germinate?
7. Discuss the relevance of different carbohydrates to the development of plant cells.
8. To what extent have enzymes been implicated in differentiation in plants?
9. Discuss what you know about the biochemistry of plant hormones.

Chapter 3

1. Give an account of the main morphological changes that accompany the development of an embryo.
2. What biochemical changes result from the fertilization of a frog or sea urchin egg?
3. In what way does early embryonic development in animals resemble germination of seeds?
4. Discuss enzymic variation during the development of sea urchin embryos.
5. Describe some biochemical changes that accompany amphibian metamorphosis.

Chapter 4

1. Compare the metabolism of carbohydrate in embryonic chicks with that in embryonic rats. How does carbohydrate metabolism alter at birth?
2. Discuss the appearance of cell-specific proteins during embryonic development.
3. Describe some systems that have been used to study induction in culture.
4. Describe some cell-specific proteins that account for the main functions of the cells in which they occur.
5. To what extent are different metabolic pathways localized in specific tissues?
6. Discuss the factors that control the rate of a metabolic pathway. Give examples of how these factors vary in differentiated cells.
7. What are isoenzymes? Illustrate their relevance to differentiation.
8. In what way is the formation of mammalian red blood cells an extreme example of differentiation?
9. Describe the distinctive features of antigens and antibodies. How far do these molecules possess cell specificity?

Chapter 5

1. What are the essential features of hormone action?
2. Describe some developmental events that are dependent on hormonally-induced enzyme synthesis.
3. Discuss the activation of enzymes by hormones.
4. What is known about receptor sites for hormones?
5. How are membranes involved in the action of hormones?
6. How far can differentiation be explained in terms of hormone action?
7. In what ways are the functions of B vitamins different from other vitamins?
8. Illustrate with examples the view that certain vitamins act like hormones.

Chapter 6

1. Discuss the relevance of cancer to differentiation. How does cancerous growth differ from organ regeneration?
2. How far has the variety of carcinogenic agents been reconciled with a single mechanism of action?
3. Discuss some relationships between carbohydrate metabolism and growth rate of cancer cells.
4. To what extent is the behaviour of cancer cells explicable in terms of enzymic alteration?
5. Describe the phenomenon of contact inhibition.
6. Assess the importance of the surface membrane in relation to the behaviour of cancer cells.

Chapter 7

1. Discuss possible mechanisms by which altered protein synthesis may be brought about during differentiation

2. Why do we believe that the genes of differentiated microbial and plant cells remain intact?
3. Describe the technique of nuclear transplantation and its relevance to differentiation.
4. Explain why genetic mutation cannot account for totipotency of differentiated cells.
5. How has nuclear transplantation been exploited to yield information regarding the site of genetic stimulation?
6. Describe the formation of hybrid cells by fusion. In what way is the function of the nucleus modified in such cells?

Chapter 8

1. Explain what is meant by transcriptional and translational control of protein synthesis.
2. What conclusions regarding protein synthesis can be drawn from experiments with enucleated cells?
3. Discuss the use of actinomycin D as an inhibitor of messenger RNA formation.
4. How has actinomycin D been used to study commitment during differentiation?
5. Discuss the appearance of specific types of RNA in development.
6. Describe the hybridization of nucleic acids and its application to the study of chromosome structure in differentiated cells.
7. What agents have been implicated as genetic repressors in higher organisms?
8. Discuss the initiation and maintenance of differentiation in animal organs.
9. What do you consider to be the outstanding biochemical problems of differentiation?

Glossary

Adaptation	*See* Section 1-1(*b*)
Allele	One of several alternative forms of a gene
Allosteric control	Control of the biological activity of proteins through an alteration in shape, caused by the binding of certain effector molecules
Amylase	Enzyme which degrades starch to smaller units
Antigen	A substance which elicits the formation of neutralizing antibody in the circulation of vertebrates
Antibody	Protein produced by vertebrates in response to stimulation by antigen
Autoradiography	Detection of radioactive material by exposure to photographic emulsion
Blastula	A hollow ball of cells formed from a fertilized egg by cleavage
Callus	Tissue that forms over a damaged plant surface
Cambial cells	Undifferentiated plant cells which give rise to secondary growth in roots and stems
Cancer	Disease due to uncontrolled growth of cells; *see* Section 6-1
Cell cycle	Sequence of events between successive cell divisions. The *S period* is that during which DNA synthesis takes place
Chloroplast	A chlorophyll-containing plant cell organelle which catalyses the process of photosynthesis
Chondrogenesis	The formation of cartilage
Cleavage	A series of cell divisions which changes a fertilized egg into a blastula
Commitment	*See* Section 1-1(*c*)
Determination	Embryological commitment to regional specialization
Differentiation	Stable development of altered structure and function
Ectoderm	The outermost layer of cells in an animal embryo
Egg	The mature female germ cell
Embryo	An organism in its earliest stage of development
Embryology	Study of formation and development of embryos
Endoderm	Inner layer of cells lining a cavity in an animal embryo
Endoplasmic reticulum	Network of double membranes within the cell cytoplasm
Epidermis	The outermost layer of cells in plants
Epithelial cells	Layer of cells covering a surface or cavity
Erythropoiesis	Production of red blood cells
Eukaryotic	Cells having a defined nucleus

Fertilization	The entry of sperm into an egg
Foetus	A vertebrate embryo
Gastrula	The cup-like structure formed by invagination of the blastula
Gastrulation	Formation of gastrula
Gene	A unit of hereditary function
Genotypic	Relating to the genetic constitution of an organism
Germination	Beginning of development in plants and certain microbes
Gluconeogenesis	Formation of glucose from non-carbohydrate precursors
Glycolysis	Degradation of glucose to lactate
Hormone	*See* Section 5-1(*a*)
Immunological technique	Use of antigen–antibody interaction to detect and separate specific substances
Immunology	Study of the formation and function of antibodies
Induction	Developmental process resulting from interaction between different cell types
Initiation	*See* Section 1-1(*d*)
Isoenzyme	An enzyme which is catalytically active in structurally different forms
Larva	A particular type of embryo
Lipase	Enzyme which hydrolyses fat
Meristem	Undifferentiated cells at growing points of plants
Mesoderm	Layer of cells lying between endoderm and ectoderm
Metamorphosis	Change of form undergone by certain animals between the embryo and adult stage
Mitosis	Division of nucleus and its contents into two halves prior to cell division
Mutation	A heritable change in genes
Neurula	A stage following gastrulation of vertebrate embryos during which the neural tube, destined to become brain and spinal cord, is formed
Notochord	Embryonic structure which in vertebrates develops into the backbone
Nucleolus	Intranuclear structure active in protein and RNA synthesis
Oogenesis	Formation of oocyte
Oocyte	Female germ cell prior to maturation into egg
Parenchyma	Relatively undifferentiated cell formed in vascular tissue of plants
Phenotypic	Relating to the properties of an organism which result from the interaction of genotypic characters with the environment
Phloem	Soft tissue forming part of the conducting system of plants

Photosynthesis	The conversion of light energy to chemical energy by plants and microbes
Placenta	The layer of tissue joining the foetal and maternal blood supply in mammals
Plasma membrane	Membrane surrounding the cytoplasm of cells
Primary growth	The elongation of plant stems and roots
Protease	Enzyme which degrades protein to smaller units
Respiration	Gaseous interchange, generally of oxygen for carbon dioxide, between cells and the surrounding medium
Ribosomes	Cytoplasmic particles composed of protein, RNA and phospholipid
Secondary growth	The thickening of plant stems and roots
Somite	Embryonic structure which develops into cartilage and muscle
Spectrophotometric analysis	Determination of chemical structures based on absorption or emission of light
Sporulation	Formation of spores
Totipotency	Ability of cell a to develop into a whole organism under appropriate conditions
Tracheid	Conducting element of xylem cells
Transcription	Synthesis of RNA from DNA template
Translation	Synthesis of protein from RNA template
Trichoblast	Plant cell from which a root hair develops
Uterus	Organ in female mammals in which embryo develops
Vitamin	Accessory food factor necessary for maximal growth of organisms
Xylem	Woody part of the conducting system of plants
Yolk	Nutrient material in eggs

Index